Movement Makes Math Meaningful

Away from the Desk Math Lessons
Aligned with the Common Core

Lisa Ann de Garcia

Part 2 – Getting Started with Math & Movement Stations | 201

Appendix:

Works Cited | 223

Preface

While working in a 90% Spanish-speaking, inner-city school in a district that was a few years into its teaching reform movement, I noticed an interesting phenomenon. There were a growing number of students who were being sent to the counseling center on a daily basis, my class included. Although there was no research study to prove it, I instinctively believed that the more rigid, structured, and somewhat scripted our teaching became, the less compatible it was for the population of students we were serving. Although our knowledge of how to teach reading, writing, and mathematics drastically improved, I could not help but wonder if we were doing a huge disservice to our students by no longer having our learning structured in such a way that provided natural movement breaks through rotations and incorporating activities that were both more engaging and meaningful in building the students' background knowledge.

One year, I felt the need to try to reach out to what I considered to be my "kinesthetic" learners. I started asking myself if there was a way to teach reading and math to my struggling 5th graders by using whole body activities. After all, I had heard the term "phycho-motor" activities from the kinder wing and knew it had something to do with moving to learn letters and sounds, but that was the extent of my understanding. Since I was stronger in my understanding of teaching mathematics, I chose to start there.

One day, on the spur of the moment, I came up with an activity using a 100-foot measuring tape. Described in more detail in part 3, it involved taking out my students and having them estimate how far 100 feet would be and to walk the entire 100 feet in increments of 10 feet, initially with their eyes open, then with their eyes closed. By the end of the 100 feet, they had internalized how far 10 feet was. Children and adults alike have such a hard time with estimating large measures, since they have such little experience measuring using those amounts. Years later, I consistently used this same activity with my university methods students and when estimating the distance of the 100 feet, I received the same results, some severely under or over-estimating. The only student who was ever spot on was a young woman who ran track.

Soon after, my teaching career took a turn and I left the classroom to become a support teacher and later a university instructor. For a few years I had forgotten about my quest of creating lessons that integrate movement to support math

learning until one day, a few years ago, I had the privilege of participating in a professional development trip to a rural school in Guatemala. On the first day, Jim Barta, the lead professor from Utah State

University, working with the 6th grade class, used masking tape to create a 100 square grid on the cement floor. He engaged the students in problems that involved computation with decimals. It was at that moment when my original question resurfaced, and this time I did not want to let it go.

Starting a new teaching job that involved pulling small groups of struggling study its affects on learning. The first year I had a few different groups of students spend about a month going around the school to measure the outside perimeters of each of the 10 buildings. We started with the rectangular buildings and then progressed to those comprised of more complicated shapes. They had to measure and record on grid paper the footprint of each building. This not only challenged their measuring skills, but their visual-spatial skills as well. One day, I was walking back to class with a group of 4th grade students when one of them spontaneously blurted, "I feel really good!" Surprised, I asked him why, to which he replied, "I don't know, but I feel really good!"

The particular student who made this comment was one that was awkward, and clumsy. I knew that something about moving and being outdoors for the 45 minutes was really good for him physically. At this point, I decided that I needed to find out what the affects of movements were on a deeper level. Was it just fun and enjoyable, or was there more to movement, which affected the body on a deeper level? This prompted me to dig into the literature and research, which has been forever life changing. What I discovered is that, not only is movement an essential modality of learning for all students, but it is absolutely critical for our struggling learners and those with special needs.

Part 1:

Introduction

Everyone in school should have a chance to learn

-Barbara Pheloung

Why Move?

It is the tradition of our education system to believe that individuals will learn best if they are presented with lots of information, in the form of a lecture or 2-dimensional written form, and seated still with eyes forward and taking notes. However, for real learning to occur, throughout our lives, hands-on learning in an environment with rich sensory experiences is optimal (Hannaford, 1995).

Even if instinctively teachers know that children need to do so, often times the classrooms are packed and teachers simply do not feel that they have the room. Others might fear chaos or a rise in discipline problems if they allow students more freedom in the classroom to move around, or simply feel that there is not enough time in the school day. However, there is plenty of evidence to support that having the children sit for long periods of time is actually doing more harm than good. In fact, it can be the very reason that discipline problems arise in the classroom in the first place.

Some movements are better than others in specifically supporting brain development. Slow, efficient, and specific movements that are designed to make sure the brain is built correctly is better than fast, disorganized movement, which is why children who are hyperactive, although always moving, still find learning difficult (Kokot, 2010). But even if teachers do know how to do this, incorporating any kind of movement in a lesson is beneficial, especially for these hyperactive learners, because they do not possess enough balance and control to sit still. Sitting still is truly uncomfortable, and their reticular activating system (RAS) of their brain needs extra stimulation of any kind to move the information on to the higher part of the cerebral cortex. Therefore, involving the senses through movement helps children pay attention and helps them recall the information by engaging the whole brain.

The research is flooding with reasons why teachers should get their students up and moving while learning new concepts. Including movement in the curriculum is beneficial because:

- ✓ It engages the whole brain in the learning
- ✓ It helps to develop social skills and increases bonding in the classroom
- ✓ It increases motivation
- ✓ It provides necessary novelty and change for the brain
- ✓ It focuses the brain to help improve concentration
- ✓ It wakes up the body when it is getting tired and keeps the brain awake and alert
- ✓ It prevents students from being overwhelmed by content
- ✓ It provides students a new perspective to the room

- ✓ It can help prevent negative effects from too much sitting
- ✓ It "awakens and activates many of our mental capacities. Movement integrates and anchors new information and experiences into our neural networks" (Hannaford 1995, p. 96)
- ✓ It has an interdependence with the body's learning systems
- ✓ It promotes an increase in test scores (Sallis, et al 1999)
- ✓ It prepares the brain for learning by getting the hemispheres to work together and increasing blood flow to the brain
- ✓ It improves cerebrospinal fluid flow
- ✓ It allows the brain to take a break and allows it time to process and consolidate information
- ✓ It allows more information to be absorbed
- ✓ It increases retention and allows students to recall the information more efficiently
- ✓ It provides a rich, multi-sensory experience for learning, which alters brain structure
- ✓ It promotes implicit learning
- ✓ It can target the root issues of disabilities such as dyslexia, dyspraxia, dyscalculia and ADHD, and increase reading, writing fluency, eye movement, cognitive skills, dexterity, and balance (Blythe 2009; Ratey 2008)
- ✓ It promotes optimal learning since the student is actively involved in exploring physical sites and materials (Gardner 1999, p. 82 cited in Jensen)
- ✓ It, when provided through distinct patterns, can supply targeted integration to weak areas of the brain
- ✓ It stimulates developing brain growth and prevents deterioration of older brains
- ✓ It creates neuropathways that connect the cerebellum to the pre-frontal cortex, which coordinates thought, attention, emotions, and social skills
- ✓ It increases communication of the neurons by myelinating brain cells causing impulses to fire faster
- ✓ It can increase the production of neurotrophins (natural nerve growth factors)
- ✓ It increases the density of the neurons in the frontal lobes, which is linked to good academic performance and executive functioning
- ✓ It balances the neurotransmitters in the brain and can stimulate the release of noradrenaline and dopamine, responsible for good feelings, therefore can act as a mood regulator and a good way to treat depression (Babyak et al 2000; Jensen 2000)
- ✓ It produces proteins that travel through the bloodstream to our brain which are responsible for making new brain cells and growing dendrites on neurons (rats that were taught complex motor skills increased their production of BDNF 35%) (Ratey 2008)

- ✓ It not only affects the motor-sensory areas of the brain, but also in the hippocampus, a structure involved in learning, which organizes, sorts, and processes the incoming information before sending it to the cortex for long-term storage (Brink, 1995)
- ✓ It creates circuits, which can be used by the thinking areas of the brain (such as the neurons grown by learning to play the piano are used for leaning math) (Ratey 2008)
- ✓ It adds an emotional context to what is being learned because all sensory stimuli passes through the limbic system, the emotional center of the brain, before proceeding to the cortex (Blomberg & Dempsey 2008)
- ✓ It releases acetylcholine across synapses of activated neurons which stimulates and attracts dendritic growth (Hannaford 1995)
- ✓ **It is fun!**

There are a variety of ways to incorporate movement into the classroom. They range from taking frequent breaks, to structuring a game or lesson that involves using the whole body, to incorporating precise movements that stimulate brain development that can help to reduce learning deficits. The purpose of this book is to not only motivate teachers to get children to move in the classroom, but to give ideas on how to incorporate it into their math lessons. It is beyond the scope of this book to explain how movements can influence changes in the brain, so for more detail regarding what is happening to the student on a physiological level while engaged in more specific movements geared towards brain development, please visit my website http://wholechildlearningsolutions.com or refer to the additional resources at the end of section 3.

The Movement-Learning Connection

> The brain is attracted to novelty
>
> The brain pays attention to movement, and
>
> The brain needs to interact with people and things in its environment
>
> -Oberparleiter, 2004, in Lengel & Kuczala, 2010, p. 19

I hold as a fundamental belief that all Children want to learn and succeed in school, although they may eventually compensate for their learning struggles by appearing not to care. To a child to whom learning does not come naturally, he has to use so much mental effort to concentrate and learn during the school day that he cannot fathom the idea of having to continue at home in the form of homework. If you have ever taken an academic class in a foreign language, you might recollect the effort it takes to concentrate. One can spend only so much time in such a focused state before all attention is lost. I can remember such a time when I took a linguistics course at a university in Mexico where I was an exchange student learning Spanish. The class was so cognitively demanding that I could only concentrate for about 15-20 minutes, after which I was really not capturing the information. My brain was simply too tired.

Children who struggle are not able to keep up with their classmates in one or more areas, which can be as frustrating to the teacher as the student. Why do so many students struggle? Many experts who have carefully observed children with learning difficulties have noticed that most, if not all, of these children also have issues with motor and balance. They have come to realize that motor development and learning go hand in hand. I have found this to be true in my own practice. A couple of years ago, when I didn't know what else to do, I simply put jump ropes in the hands of my 4th -6th grade students and not a one of them initially could jump rope. Now, I not only test my students for his developmental level in mathematics (see Appendix D for my developmental math assessment), but I check for a variety of markers of their motor development as well. I have been noticing that students with more severe learning problems also have more severe motor issues than others.

Learning difficulties are often neurologically based, and can also lead to behavior and emotional problems. Studies have indicated that more than 80% of prisoners had a serious learning problem as a child (Ratey, 2008). Allan Bermann found that visual perception was the disability that occurred most often in a group of delinquent children, followed by auditory memory and language deficit (Phelong, 1997). If we rather than just an academic or behavioral one, especially at a young age, how many of them could we save from a lifetime of struggle?

Addressing learning issues on the academic level is like repairing a roof when the walls and foundations are cracked and crumbling. The following diagram, *The Learning Ladder,* illustrates what systems need to be in place, and in what order, so that they are able to appropriately support academic learning, located at the top of the ladder.

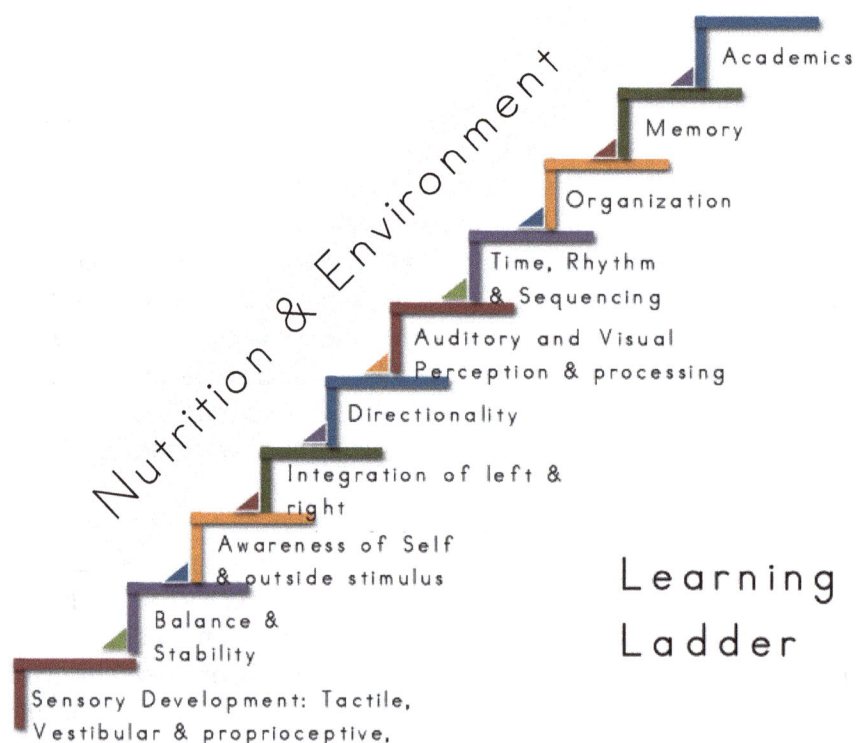

Academics

Memory

Organization

Time, Rhythm & Sequencing

Auditory and Visual Perception & processing

Directionality

Integration of left & right

Awareness of Self & outside stimulus

Balance & Stability

Sensory Development: Tactile, Vestibular & proprioceptive,

Nutrition & Environment

Learning Ladder

The Learning Ladder – how we developmentally learn (de Garcia 2014).

When we look at babies, we see that they do not yet have the neural connections into the frontal lobes of their brain to control their impulses. It is normal, at this age, for babies to be hyperactive because certain parts of the brain, the basal ganglia in particular, have not yet been developed and are not connected to other levels of the brain (Blomberg & Dempsey, 2011). Children who hop, spin and crash into walls while walking are still learning to control their balance. They are demonstrating that they too have underdeveloped brains and are developmentally similar in some ways to the active infant.

What these children are silently telling us is that somewhere along the line, they have missed some critical developmental stages, because the brain does not develop normally if a stage of development is missed (Gold, 2008). When analyzing the behaviors of poor readers, the problems that had been identified all boiled down to an unorganized nervous system (Gold, 2008). Eye dominance is one result of this organization. Studies have showed that as much as 81% of students who have learning difficulties are left-eyed and right handed. "Since the eye naturally wants to track from the right to left, it will also guide the hand from the right to left, which may cause writing difficulties or letter reversals" (Hannaford, 1995, p. 211).

Sally Goddard Blythe, in her book *The Well Balanced Child* explains that many of the symptoms that are expressed in disorders, such as dyslexia, attention deficit, or anxiety disorders are actually caused by a "treatable signal-scrambling dysfunction" within the inner ear and cerebellum (Goddard, 2007). These are the same symptoms that can be "triggered in normal individuals following excessive spinning and dizziness." Studies have found that participating in martial arts, due to its demand on focus in combination with aerobic activity, twice a week improves behavior and performance of children suffering from this disorder.

Someone with dyspraxia has a disorganization of movement, particularly unfamiliar movement and those involving multiple steps. Deficit in motor planning and sequencing is often a leading factor in a variety of developmental and motor deficits, including speech. Motor development progresses from head to toes and from the core outward. Therefore, there may be little connection to the feet, although the upper body may appear well-coordinated. Dr. Blomberg (2011) explains that movement ability and speech are linked and that stimulating the cerebellum to improve motor abilities need to happen before speech can be improved.

Dyscalculia, a disorder in calculation, is defined by the National Center for Learning Disabilitites (2006) as "a wide range of life long disabilities involving math." Specific areas in the left hemisphere used in counting, calculating, and using basic arithmetic number symbols are located mostly in regions of the left parietal lobe and motor cortex. Areas in the prefrontal cortex are used in analyzing a problem and retrieval of facts. Regions in the right parietal lobe are used in spatial reasoning and visual-spatial tasks, like being able to generate a mental number line, and estimating.

Students with dyscalculia have significant weaknesses in areas on the left hemisphere that effect their ability to compute or recall basic facts (Sousa 2008). They may equally have difficulty in reading, or dyslexia, since decoding and phonemic awareness are also located on the left side.

Although not as commonly known as dyslexia, there is actually a significant number of students, between 20% - 60%, who have both (Butterworth & Yeo, 2004; Hannell, 2005). This means that many students with language related issues struggle with math as well, and that these students also experience motor skill deficits. Their right side areas are

generally functioning properly or may even be well above average. If these children are not in an environment that embraces understanding and conceptual thinking, they will have limited access to understanding mathematics, even though they may very well be destined to be great mathematical thinkers. Movement helps encourage the use of both sides of the brain during math, increasing assess of the weaker side and communication between both hemispheres. Sue Peace in the UK, and expert in dyscalculia, has designed a course through Brain Gym™ entitled *Kinesiology 4 Numeracy.*

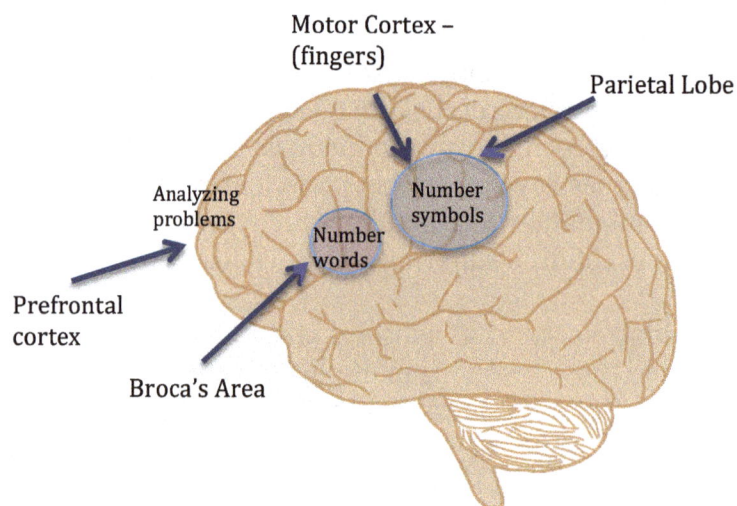

Source: *How the Brain Learns Mathematics,* by David Sousa (2008)

The brain of a child with autism is highly aroused and not comfortable in its own resting state (Othmer, 2012). One theory is that there is an overproduction of BDNF, the growth factor that helps neurons to wire together. Areas of the cerebellum, on the on the right hemisphere, which are linked to the speech areas of the left hemisphere, tend to be smaller than normal in children with autism (Blomberg & Dempsey, 2011). There is also an area on the brainstem, which visual control, vestibular input, and proprioceptive information come together, when not functioning properly causes paralysis of gaze, common in children with autism (Gold, 2008).

World-known researchers and practitioners, such as Sally Goddard-Blythe, Dr. Harald Blomberg, and Svetlana Masgutova, just to name a few, have spent many years investigating the role movement has in neurological development and learning. According to Goddard-Blythe (2005), "attention, balance and coordination are the primary A, B, and C upon which all later academic learning depends" (p. xvi). Reading, for example, depends upon smooth eye movements across the page, which is developed by the balance system. These, and others, have found that children labeled as "learning disabled" were able to more effectively learn when they spent a few minutes before a lesson with simple, whole body integrative movements. This is why Brain Gym™ is so successful. It is a way to help the body prepare for learning.

Shirley Kokot (2010) explains that body movements are responsible for the development of the brain structure and they contribute to the proper functioning of the brain. Roger Spery, a Nobel Laureate neurobiologist said that, "90% of stimulation and nutrition to the brain is generated by movement of the spine."

Movement is involved in each of the senses. For example, sound moves in sound waves, touch is perceived by movement air or pressure over the skin. Movement helps by enhancing functioning of the nervous system and causes chemicals in the brain to be created that allow the neurons to communicate with one another and to make general processing faster (Kokot, 2010).

When you observe a baby, you will notice that when awake, it is never still. The movements of the infant are rapidly growing the brain. It is estimated that in the first year of life, every minute there are more than 4 million new nerve cell branches created in the brain (Blomberg & Dempsey, 2011). Babies who are, for whatever reason, unable to move much are certain to have developmental delays. These children need to be moved passively, such as by rocking and touch, to stimulate their brains (Blomberg & Dempsey, 2011).

These examples demonstrate how critically interconnected movement and learning are. It is no wonder that in this modern day of computer and video games that there is such a surge of students with learning and attention difficulties. If young children are no longer spending hours on end outdoors exploring and working on developing and integrating their sensory systems, then I argue that it is incumbent upon us, as educators, to provide supplemental experiences that allow them to do so.

Children, and many adults, have short attention spans and need to have frequent learning breaks. The average attention span for a child is said to be their age +2 minutes. That means after that many minutes, teachers need to stop for a break to allow children to process the information. Elementary schools in Japan teach for 50 minutes and provide a 15 minute recess after each 50 minute learning session. Schools in Finland now recognize the value of periodic movement for both students and teachers. As adults, we typically will push through the day working, thinking

and preparing during all of our breaks. Rest is just as important for us as for the young ones.

The lessons in this book are designed to benefit all students, not just those that struggle. Movement helps everyone access their whole brain while learning, and besides, it is fun and motivating. Students need to move before they are able to sit still, so if it feels like the classroom is getting out of control and everyone is tuning you out, get them up and (with well-defined parameters) move!

How to read this book

The pages in this text are full of ideas that can be applied immediately to the classroom.

Section two is composed of more than 50 math lessons, aligned with the common core. They are organized based on the specific equipment or tools needed, most of which are commonly found in a typical classroom. These lessons may not incorporate the precise brain-developing movements of section two, but the purposes and of these movements are just as important. Each lesson has a "Did you Know" section which provide bite-size pieces of information on movement and its relation to brain structure and the theories of neuroplasticity. Lessons also either have a section entitled "Make it Memorable" which give advice for supporting memory, or "A Teacher's Insight" box, which, is exactly as it sounds, advice a teacher has regarding either a particular lesson or teaching concept.

Section three begins by giving the reader a brief snapshot of our senses and how they relate to learning. The purpose is to introduce the reader to the role the sensory system plays in learning and to spark an interest to do further reading and study on the subject. Then it describes setting up movement stations that embed mathematics throughout. The movements are designed to promote brain development by strengthening the vestibular, visual, balance, and other systems. These stations might either be set up in a vacant room to be shared by other classes, or cleverly set up in the classroom itself.

Finally, in **Appendix A** one will find a Common Core overview by grade and **Appendix B** provides a quick reference chart of the lessons sorted by common core standards.

Unfortunately, teacher education programs do not talk about the role movement plays in learning, therefore, many still promote a traditional-style curriculum and value when children are quietly sitting and listening to their teacher, although sitting still can be nearly impossible for some. Therefore, it is the responsibility of all who reads this book to continue researching, share it with your colleagues, observe children and learn together.

Part 2:

Classroom Lessons that Integrate Math and Movement

Movement starts from Experience, not just looking at a book.

-Barbara Pheloung

The following lessons are organized by materials needed, and are correlated with standards from the Common Core. These activities will help get the children up and moving, enhance engagement, and help them interact with themselves and their environment. They also help students to be able to access both sides of their brain while learning, which in turn helps to reduce stress, promotes a better sense of well being, and improves visual-motor skills, just to name a few.

The Giant 100 Grid

I have discovered that a 100 chart, blank or filled in, can be used for so many lessons from kindergarten through 6th grade. I also discovered how easy it is to blow it up and let students solve problems on it together. Usually, I will have a small group working together on the mat and the rest of the class simultaneously solves problems with a small 100 chart, rotating students for each problem.

Giant 100 Grid Lessons

Lesson Name	Grade Level	CCSSI Standard
Building the 100 Chart	1st-3rd	CCSS.Math.Practice.MP5 CCSS.Math.Practice.MP7
Add 10	1st & 2nd	CCSS.Math.Content.1.NBT.C.5 CCSS.Math.Content.2.NBT.B.8
Connecting the Number Line to the 100 Chart	1st & 2nd	CCSS.Math.Practice.MP4
Skip Counting Patterns	1st-3rd	CCSS.Math.Content.3.OA.D.9
Addition with Regrouping	2nd	CCSS.Math.Content.2.NBT.B.5
Subtraction with Regrouping	2nd	CCSS.Math.Content.2.NBT.B.5
Addition with Decimals	5th	CCSS.Math.Content.5.NBT.B.7
Subtraction with Decimals	5th	CCSS.Math.Content.5.NBT.B.7
Building the Multiplication Chart	3rd	CCSS.Math.Content.3.OA.A.1 CCSS.Math.Content.3.OA.B.5 CCSS.Math.Content.3.OA.C.7
Perimeter	3rd	CCSS.Math.Content.3.MD.D.8
Coordinate Graphing	5th & 6th	CCSS.Math.Content.5.G.A.1 CCSS.Math.Content.5.G.A.2 CCSS.Math.Content.6.NS.C.8
Experimental vs. Theoretical Probability	7th	CCSS.Math.Content.7.SP.C.7 CCSS.Math.Content.7.SP.C.7a

Note: Grade level indicates where they appear in the Common Core or is typical in curriculum, however, it may be appropriate to do a lesson in other grade levels depending on conceptual understanding.

Making the Giant 100 Grid

1- Buy clear plastic used for tablecloth at a store such as Wal-Mart. You will find it in the fabric section next to the measuring counter. You will need 4 times the width.

2- Lay out the plastic on a large flat surface and cut it exactly in half.

3- Place the two lengths next to each other and you should have a square. Use clear packaging tape to attach the two halves.

4 – Using a measuring tape measure one length of your square. Calculate how much will be 1/10 and mark tenths with a pen on all sides of the square.

5- Using painter's tape, and preferably two people, lay the tape from one mark to the one opposite to it.

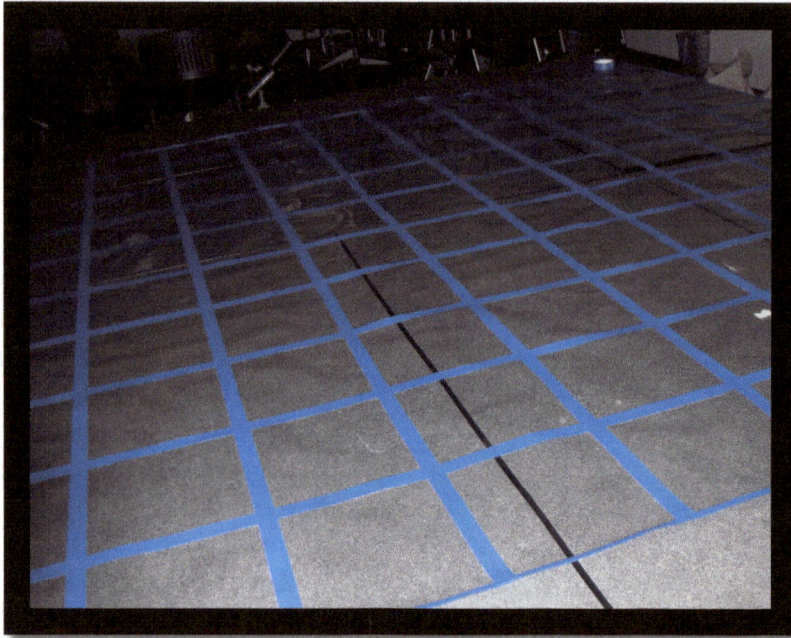

6 – when you are through, it should look like the picture to the right.

7 – Enjoy!

My methods students were so excited about the 100 grid that they came in on a Saturday to create some for themselves.

Building the 100 Chart

Grade Level: 1-3

Materials: Giant 100 grid, dry erase markers, towels or something to use as erasers

Common Core Connection:
(Standards for Mathematical Practice)

CCSS.Math.Practice.MP5 Use appropriate tools strategically

CCSS.Math.Practice.MP7 Look for and make use of structure

Objective: In the context of working together to fill out the 100 grid, students will explore patterns that are inherent within the 100 chart.

Introduction	Present a group of up to about 5 students with the blank giant 100 grid, and a dry erase marker for each one. Instruct them that they will be filling it out to make a 100 chart. Make sure that there is no 100 chart visible in the classroom. Do not guide them much in this activity, but do tell them which square is the "1".
Exploration	Students fill out the 100 chart. Having multiple students will cause them to try to split the work. This is more problematic than if there is just one student doing it alone. What will probably happen is that they will not have planned it out very well together and they will end up with some inconsistencies and they will have to re-do some of their work. Try to wait and let them discover the problem areas and work it out for themselves.
Discussion	Ask questions based on what you noticed the problem areas to be. Notice if they caught onto a pattern when they fill out the chart, such as write in all the tens first. You might ask questions such as, "what was challenging about this task?" "What made you decide to....?" "What patterns do you notice?"

Note: I have used this lesson with struggling 4th & 5th graders to help them make sense of the structure of the 100 grid and tens and ones.

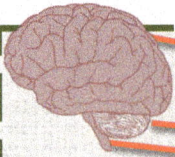

There are three main parts of the brain: **brainstem**, **cerebellum**, and **cortex**, which are involved in learning.

The **brainstem** regulates unconscious motor activity, regulates sense of balance, eye movement, visual perception, auditory processing, and relays information from sensory organs.

The **Cerebellum** helps the body maintain a sense of balance and coordinates muscular movements with sensory information

The **cortex**, designed to handle abstract thought and higher-order thinking, has to work overtime to carry out responsibilities that the brain stem, which regulates unconscious motor activity (such as eye movement) has not been trained to handle.

The brain develops from the inside out. If the inner parts, brainstem and cerebellum, are not properly developed, the cortex cannot be fully connected.

Make it Memorable

Memory includes visual memory, verbal/auditory memory, and working memory. Some signs that students are struggling in one of these areas are:

- ❖ Inability to retain math facts or new information
- ❖ Forgetting steps in an algorithm
- ❖ Performing poorly on review lessons
- ❖ Difficulty telling time
- ❖ Difficulty solving multi-step word problems

Adding 10

Grade Level: Grade 1 & Grade 2

Materials: Giant 100 grid that is filled in with the numbers 1-100, (either with dry erase marker, post-it notes, or index cards taped to the back of the clear plastic so that the numbers are showing through), individual hundred charts, or a class hundred pocket chart.

Common Core Connection:

Grade 1
CCSS.Math.Content.1.NBT.C.5 Given a two-digit number, mentally find 10 more or 10 less than the number, without having to count; reasoning used.

Grade 2
CCSS.Math.Content.2.NBT.B.8 Mentally 100 to a given number 100–900, and mentally subtract 10 or 100 from a given number 100–900.

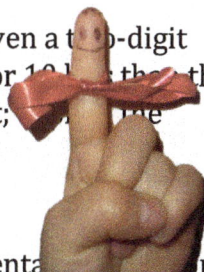

Objective: In the context of exploring addition of 10 on a giant hundred grid, students will begin to internalize that when adding 10, the answer is below the original number.

Introduction	The teacher asks a student to stand on a given square on the giant 100 grid. The teacher then asks the class if they were to add ten more squares, where would they be. Another student is then asked to come up and stand where 10 more would be and to justify where he chose to stand. The rest of the students are using their individual hundred charts and using counters.
	Note: it is very common that the student will have started on the square next to the original student, walked to the end of that row and then doubled back onto the next row starting from ten. For example:

1	2	3	4	5	6	7	8	9	10
11	12	13	14	15	16	17	18	19	20
21	22	23	24	25	26	27	28	29	30
31	32	33	34	35	36	37	38	39	40

If this happens, the rest of the class need to decide if they agree with his strategy, since they will have been solving the problem on their individual hundred chart. The class needs to come to a consensus that when arriving at the ten, in this case 30, one needs to continue by going to the beginning of the next row.

Once the students are standing on their correct spots, then another student is asked to find the number that will be 10 more than the previous. After a few students, the grid should look like:

1	2	3	4	5	6	7	8	9	10
11	12	13	14	15	16	17	18	19	20
21	22	23	24	25	26	27	28	29	30
31	32	33	34	35	36	37	38	39	40
41	42	43	44	45	46	47	48	49	50
51	52	53	54	55	56	57	58	59	60
61	62	63	64	65	66	67	68	69	70

The teacher needs to ask the class what they notice. They should notice that all the people, or counters, are in the same row. The teacher should ask the students if they think that will always happen and why or why not.

Exploration	In partnerships or groups of 3, students are asked to explore what they noticed with other numbers to see if it will always happen and why. Students are given the opportunity to choose to use the giant 100 chart, the class 100 pocket chart, or their individual charts. This will allow students who need the movement to select it as their tool. When exploring, students should fill out a recording tool (see next page) to keep track of their thinking and to notice patterns. Have students color in the squares on their recording tool.
Discussion	As a whole class, sitting around the giant 100 grid, students should discuss what they notice. The teacher should invite students to give an example of what they found by demonstrating on the grid. By the end of this session, students begin to articulate that when adding 10, the answer is below the original number.

Next steps:

Once students have generalized +10, they can engage in similar activities for -10, +9, -9, +20, +30, etc. Using the hundred chart as a mental frame will help first and second graders compute mentally.

Adding 10 Recording Tool

1. We chose the number _____

   When we added 10, we landed on _____

   When we added 10 more, we landed on _____

   When we added 10 more, we landed on _____

1	2	3	4	5	6	7	8	9	10
11	12	13	14	15	16	17	18	19	20
21	22	23	24	25	26	27	28	29	30
31	32	33	34	35	36	37	38	39	40
41	42	43	44	45	46	47	48	49	50
51	52	53	54	55	56	57	58	59	60
61	62	63	64	65	66	67	68	69	70
71	72	73	74	75	76	77	78	79	80
81	82	83	84	85	86	87	88	89	90
91	92	93	94	95	96	97	98	99	100

2. We chose the number _____

   When we added 10, we landed on _____

   When we added 10 more, we landed on _____

   When we added 10 more, we landed on _____

1	2	3	4	5	6	7	8	9	10
11	12	13	14	15	16	17	18	19	20
21	22	23	24	25	26	27	28	29	30
31	32	33	34	35	36	37	38	39	40
41	42	43	44	45	46	47	48	49	50
51	52	53	54	55	56	57	58	59	60
61	62	63	64	65	66	67	68	69	70
71	72	73	74	75	76	77	78	79	80
81	82	83	84	85	86	87	88	89	90
91	92	93	94	95	96	97	98	99	100

3. We chose the number _____

   When we added 10, we landed on _____

   When we added 10 more, we landed on _____

   When we added 10 more, we landed on _____

1	2	3	4	5	6	7	8	9	10
11	12	13	14	15	16	17	18	19	20
21	22	23	24	25	26	27	28	29	30
31	32	33	34	35	36	37	38	39	40
41	42	43	44	45	46	47	48	49	50
51	52	53	54	55	56	57	58	59	60
61	62	63	64	65	66	67	68	69	70
71	72	73	74	75	76	77	78	79	80
81	82	83	84	85	86	87	88	89	90
91	92	93	94	95	96	97	98	99	100

Adding 10 Recording Tool

1. We chose the number _____

    When we added 10, we landed on _____

    When we added 10 more, we landed on _____

    When we added 10 more, we landed on _____

1	2	3	4	5	6	7	8	9	10
11	12	13	14	15	16	17	18	19	20
21	22	23	24	25	26	27	28	29	30
31	32	33	34	35	36	37	38	39	40
41	42	43	44	45	46	47	48	49	50
51	52	53	54	55	56	57	58	59	60
61	62	63	64	65	66	67	68	69	70
71	72	73	74	75	76	77	78	79	80
81	82	83	84	85	86	87	88	89	90
91	92	93	94	95	96	97	98	99	100

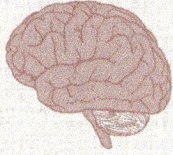

A Teacher's Insight

The idea that 10+ a number is just underneath the number on the 100 chart is an idea that needs to be explored by students again and again before it is internalized and trusted that it always works. Just like when constructing tens with base 10 blocks, some students may need to count out 10 many times before seeing the relationship. It may frustrate teachers that after students simply do not "see" that they can just move to the umber below, but like any mathematical concept, this relationship must be constructed by the learner. It is the job of the teacher to provide the appropriate experiences to help children to see and articulate this relationship.

Connecting the Number line to the 100 Chart

Grade Level: Grade 1 & Grade 2

Materials: Giant 100 grid, 100 pieces of paper that are each cut to the size of the individual squares in the grid, tape, markers.

Common Core Connection:

CCSS.Math.Practice.MP4 Model with mathematics.

Objective: In the context of creating a giant number line and cutting it into strips of 10, students will make connections between the number line and hundred chart as tools for counting and adding.

Introduction	The teacher should ask the students if they think there is a relationship between a number line and hundred chart (assuming that they have been using both in class) and what that relationship might be.
Exploration	Provide each small group of children (3-4) with enough square pieces of paper so that there are 100 pieces between them. Assign the students certain numbers that they will write on each paper (in large letter). Students should go out into the hall, auditorium, or playground and assemble their numbers from 1-100. They should them tape the numbers together. Tell the students that since the number line is too long, they are going to organize it in a different way so they can put it on their wall in the classroom. Ask for suggestions on how to break it up. If no one suggests breaking into tens, make that a suggestion based off of some work you have done in class. Allow students to cut the number line into segments of ten and ask them how they can arrange it so it won't be so long. Someone should suggest to put them on top of one another. Once they are all arranged, tape them together. The newly formed 100 chart can be laid out on the giant 100 grid (or underneath so the numbers show through) and used for future activities.

Discussion	Have students discuss what they notice. They will say that it is a 100 chart. Facilitate a discussion where the children are noticing the relationships between each of the tools and, although organized differently, the two are related.

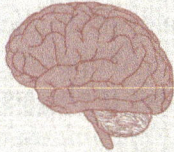

Did you know?   #3

Paul MacLean describes the brain as being in layers like an onion. The most inner part of the brain is the **brain stem**, commonly known as the "fish brain." The function of this part of the brain is to receive signals from our senses and to relay them to the motor organs. All of our automatic functions are controlled by the brain stem.

Make it Memorable

Help Students connect new material with movement to make retaining ad recalling the information easier by:

Performing windmills or "Head, Shoulders, Knees and Toes" while counting or reciting the alphabet

Skip Counting Patterns

Grade Level: 1-3

Materials: Giant 100 Grid, index cards, paper 100 charts (one for every number you plan to skip count), crayons or markers.

Common Core Connection:

Grade 3
CCSS.Math.Content.3.OA.D.9 Identify arithmetic patterns (including patterns in the addition table or multiplication table), and explain them using properties of operations. *For example, observe that 4 times a number is always even, and explain why 4 times a number can be decomposed into two equal addends.*

Objective: In the context of creating skip counting patterns on the 100 grid, students will notice patterns that are made and use them to help learn how to skip count.

Introduction	Ask students to look at the 100 grid and visualize what pattern would be made if they put an index card on all the 2's. After some suggestions, provide a couple of students with a stack of index cards and ask them to start on one and to skip count 2's by walking on each number they say and to lay an index card on that number. When they finish, ask them to look at the pattern created and describe what they notice. Ask students whether or not this was the pattern that they had envisioned. Ask the class to record what was made onto their paper 100 charts using a crayon or marker.
Exploration	Now, ask students to do the same for 3, 4, 5, 6, etc. Have them first visualize and then mark the spots on the grid, allowing students to take turns. Students should then mark their papers in the same pattern created on the grid (or they can mark their papers and then compared to the finished mat). In small groups, allow students to compare the different numbers and find patterns, similarities and differences
Discussion	As a large group, ask questions such as, "what is the pattern made on the chart when skip counting by 3's? Is there any other number that makes a similar pattern? Why? How can using a hundred chart help us learn our skip counting?

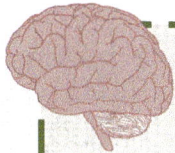

Did you know? #4

The basal Ganglia, part of the brainstem, is "responsible for the organization of involuntary and semi-voluntary activity, upon which consciously willed movements are superimposed" (Goddard, 2005, p. 44). It "connects and orchestrates impulses between the cerebellum and frontal lobe, thus helping to control body movement" (Hannaford, 1995, p. 60).

A Teacher's Insight

Allowing students to make a pattern on the grid while moving helps them to feel the pattern as they are walking and skipping so many squares each time. This helps when they step back to observe to recognize the pattern with their eyes.

Addition with Regrouping

Grade Level: Grade 2

Materials: Giant 100 grid (blank); about 30 base-ten 100 flats, strips of register tape (the size of the length of the mat).

Objective: In the context of adding on a hundred chart, students will begin to generate strategies for regrouping which they can apply to a pictorial and numerical model.

<table>
<tr>
<td>Common Core Connection:

Grade 2
CCSS.Math.Content.2.NBT.B.5 Fluently add and subtract within 100 using strategies based on place value, properties of operations, and/or the relationship between addition and subtraction.</td>
</tr>
</table>

Introduction	The teacher asks students to represent a number on the 100 grid (such as 18) with the materials provided.
	➤ Students might choose to either lay out 18 flats or use the pieces of register tape to represent 10s. Then ask them to represent a second number, while the first one is still present (such as 21). Ask the group how many they have altogether and how they know.
	➤ Students should be saying things like: putting the tens together and the ones together to get 39. If they haven't actually re-organized the pieces to do that, then encourage them to do so. Students sometimes are just mentally organizing them and giving an answer.

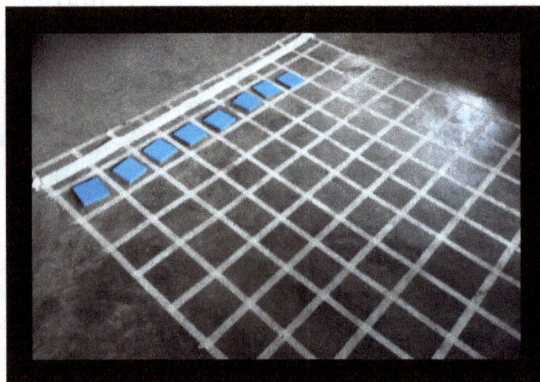

Exploration	Provide the next problem, 38 + 24 and ask them to solve using their tools. ➤ Watch to see if they just add 4 more, or if they scrap the entire problem and start all over. Before moving on, discuss what they did and why. Notice if they converted the 10 flats to a strip or if they just counted the tens and then counted on the 12 ones. If they did not trade, ask them what they notice with what they have made and guide them to notice their new ten and ask them how they could now represent that. Provide additional problems to work on as a group such as: 38+25; 45+27; and 62+19.
Discussion	After students solve each problem, ask questions such as, "How did you combine the numbers?" "How do you know when you need to trade?" "What number are you breaking up to make a ten?" After several problems, students should be using making ten strategies for combining their ones together. For example, when adding 19+25, they might take one from the 5 and put it with the 9 to make ten and 4 left over.

➤ Students should begin solving similar problems using base ten blocks and directly modeling problems the same way they did as a class. For additional support, blank 100 charts (10 x 10 cm) may be needed.

➤ Students need to record in pictures how they are solving with the blocks. See the example below for 27 + 38:

$$20 + 30 = 50$$
$$7 + 3 = 10$$
$$50 + 10 + 5 = 65$$

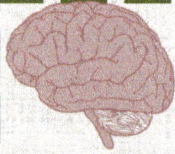

Did you know?

The brain stem also has a net of nerve cells called the Reticular Activating System (RAS). The job of the RAS is to receive impulses from all our senses, except for the sense of smell, and then to transmit them to the cortex, which improves attention and alertness. If the cortex is insufficiently stimulated by the RAS, then the child will be passive and will be unable to pay attention.

Make it Memorable

Help Students connect new material with movement to make retaining ad recalling the information easier by:

Juggling one or two balls , bean bags, or
scarves while counting odd and even numbers

Subtraction with Regrouping

Grade Level: Grade 2

Materials: Giant 100 grid (blank); about 30 base-ten 100 flats, strips of register tape (the size of the length of the mat).

Objective: In the context of subtracting on a hundred chart, students will begin to generate strategies for regrouping which they can apply to a pictorial and numerical model.

Common Core Connection:

Grade 2
CCSS.Math.Content.2.NBT.B.5 Fluently add and subtract within 100 using strategies based on place value, properties of operations, and/or the relationship between addition and subtraction..

Introduction	The teacher asks students to represent a number on the 100 grid, such as 39, with the materials provided.
	➢ Students might choose to either lay out 39 flats or use the pieces of register tape to represent 10s. If they use 39 flats, ask them how else they could represent 39 with less pieces.
	Ask how many they have if they take 17 away.
	➢ Students will most likely simply remove a ten and seven ones. Ask what number sentence that represents and write it on the board.
Exploration	Now ask students what if they only had 32 (changing the problem that is written on the board). Ask students to represent 32 and then attempt to remove 17.
	➢ Most students will simply remove one strip away, but may be perplexed as to how to take the extra 7. They might take 2 away, since there are two available, but not know how to take away the remaining 5, therefore may say that the answer is 20. Reminding them that they needed to take away 7 and not 2, ask them for other suggestions. Some may suggest to rip the paper, but you should tell them that you need the strips for other problems, and if there is another way that they might be able to take away without ripping or bending the paper.

	If no one comes up with a solution, ask how much the paper is worth (to which they will reply "10") and ask if there is a different way to represent 10 with the other tools that are available. By now, there will certainly be some "ah-ha's" and students will trade out the strip of paper for ten flats, thus allowing them to remove the remaining amount.
	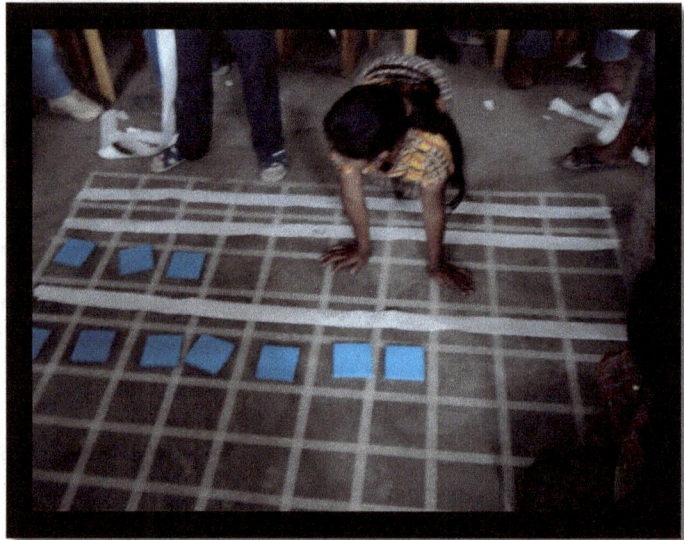
	Provide the next problem, 45 - 27 and ask them to solve using their tools. Before moving on, discuss what they did and why. Notice if they converted the now 10 flats into a strip. Provide additional problems to work on as a group such as: 25-18; and 62-19. Notice if they are trading and then removing the amount, or if they are mentally counting how many they would have left after subtracting, then removing the strip and just putting what is left. For example, if they need to remove 8 and there are only 5 available, do they trade one of the tens to now have 15 or do they see that if they take the 5 away, they will need to remove 3 more and then look to the ten to see that if they remove 3 from that ten 7 would be left.
Discussion	After students solve each problem, ask questions such as, "What did you remove first?" "Why?" "Why did you need to trade the ten?" "How would you know when you would need to trade or not?"

> Students should begin solving similar problems using base ten blocks and directly modeling problems the same way they did as a class. For additional support, blank 100 charts (10 x 10 cm) may be needed.

> Students should be recording in pictures how they are solving with their base-ten blocks. See the example below for 72-46:

In this example, the student is showing that he traded a ten for ones and removed the 4 tens and 6 ones.

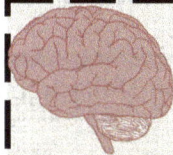

A Teacher's Insight

One of the most important concepts for teaching, in my opinion, is the notion of helping children move from the concrete to abstract representation in an appropriate way and pace. One of the most concrete forms of representation is acting out the problem literally. Having students take an abstract procedure and representing it in a more concrete way, like acting out with blocks, can help them build understanding of that procedure so they can understand what all the steps actually represent.

Addition with Decimals

Grade Level: 5th Grade

Materials: Giant 100 grid (blank); about 30 base-ten 100 flats, strips of register tape (the size of the length of the mat).

Objective: In the context of adding on a hundred chart, students will begin to generate strategies for regrouping of decimals which they can apply to a pictorial and numerical model.

Common Core Connection:

Grade 5

CCSS.Math.Content.5.NBT.B.7 Add, subtract, multiply, and divide decimals to hundredths, using concrete models or drawings and strategies based on place value, properties of operations, and/or the relationship between addition and subtraction; relate the strategy to a written method and explain the reasoning used.

---

*Note: this lesson is similar to the Addition with regrouping lesson, but now the grid is one whole and the amount represented is in hundredths.

Introduction	The teacher asks students to represent a number on the 100 grid (such as .18) with the materials provided.
	➤ Students might choose to either lay out 18 flats or use the pieces of register tape to represent tenths. Then ask them to represent a second number, while the first one is still present (such as .21). Ask the group how much they have altogether and how they know. 
	➤ Students will say things like: "we are putting the tenths together and the hundredths together to get .39." If they haven't actually re-organized the pieces to do that, then encourage them to do so. Students sometime just mentally organize them and giving an answer.

Exploration	Provide the next problem, .38 + .24 and ask them to solve using their tools. ➤ Watch to see if they just add 4 more flats, or if they scrap the entire problem and start all over. Before moving on, discuss what they did and why. Notice if they converted the 10 flats to a strip or if they just counted the tenths and then counted on the 12 hundredths. If they did not trade, ask them what they notice with what they have made and guide them to notice their new ten and ask them how they could now represent that. Provide additional problems to work on as a group such as: .38+.25; .45+.27; and .62+.19.
Discussion	After students solve each problem, ask questions such as, "What is the whole?" "How much does the strip represent?" "How much does the flat represent." "How did you combine the numbers?" "How do you know when you need to trade?" "What number are you breaking up to make a tenth?" After several problems, students will begin using making ten strategies for combining their ones together. For example, when adding .19+.25, they would take one from the 5 and put it with the 9 to make a tenth and 4 hundredths left over.

➤ Students need to begin solving similar problems using base ten blocks and directly modeling problems the same way they did as a class. For additional support, blank 100 charts (10 x 10 cm) may be needed.

➤ As students move toward adding numbers larger than one (for example, 2.45), students should be provided with multiple 10 x 10 cm grids so they can directly model such problems.

➤ Ask students how much the small cube on the flat would represent in comparison to the entire grid and in comparison to the strip of paper.

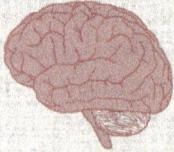

The **cerebellum,** which contains ½ of the brain's neurons, receives signals from receptors for the kinesthetic and tactile senses that transmit information regarding touch and pressure (Blomberg & Dempsey, 2011). It is involved in various aspects of planning and monitoring movements and regulates muscle tone, including saccadic eye movements. Its job is to make our movements coordinated and smooth. Apart from motor control, it also is involved in attention, long-term memory, spatial perception, impulse control, abstract thinking and other cognitive functions (Lengel & Kuczala, 2010), therefore, movement has a direct affect on the latter, including eye movements, reading comprehension, speed of information processing, working memory, learning and speech development.

Make it Memorable

Help Students connect new material with movement to make retaining ad recalling the information easier by:

Pointing to designated locations around the room while Holding a balance pose, such as north, south, east, or west.

Subtraction with Decimals

Grade Level: 5th Grade

Materials: Giant 100 grid (blank); about 30 base-ten 100 flats, strips of register tape (the size of the length of the mat).

Objective: In the context of subtracting on a hundred chart, students will begin to generate strategies for regrouping of decimals which they can apply to a pictorial and numerical model.

*Note: this lesson is similar to the Addition with regrouping lesson, but now the grid is one whole and the amount represented is in hundredths.

Common Core Connection:

Grade 5
Perform operations with multi-digit whole numbers and with decimals to hundredths CCSS.Math.Content.5.NBT.B.7 Add, subtract, multiply, and divide decimals to hundredths, using concrete models or drawings and strategies based on place value, properties of operations, and/or the relationship between addition and subtraction; relate the strategy to a written method and explain the reasoning used.

Introduction	The teacher asks students to represent a number on the 100 grid, such as .39, with the materials provided.
	➢ Students might choose to either lay out 39 flats or use the pieces of register tape to represent tenths. If they use 39 flats, ask them how else they could represent 39 with less pieces.
	Ask how much they have if they take .17 away.
	➢ Students will most likely simply remove a tenth and seven hundredths. Ask what number sentence that represents and write it on the board.
Exploration	Now ask students what if they only had .32 (changing the problem that is written on the board). Ask students to represent .32 and then attempt to remove .17.
	➢ Students will probably remove one strip away, but may be perplexed as to how to take the extra .07. They might take 2 flats away, since there are two available, but not know how to take away the remaining 5, therefore may say that the answer is .20.

	Reminding them that they needed to take away .07 and not .02, ask them for other suggestions. Some may suggest to rip the paper, but tell them that you need the strips for other problems, and if there is another way that they might be able to take away without ripping or bending the paper. If no one comes up with a solution, ask how much the paper is worth (to which they will reply ".1") and ask if there is a different way to represent .1 with the other tools that are available. By now, there will certainly be some "ah-ha's" and students will trade out the strip of paper for ten flats, thus allowing them to remove the remaining amount. Provide the next problem, .45 - .27 and ask them to solve using their tools. Before moving on, discuss what they did and why. Notice if they converted the 10 flats into a strip. Provide additional problems to work on as a group such as: .25 - .18; and .62 - .19. ➤ Notice if they are trading and then removing the amount, or if they are mentally counting how many they would have left after subtracting, then removing the strip and just putting what is left. For example, if they need to remove 8 and there are only 5 available, do they trade one of the tenths to now have 15 flats or do they see that if they take the 5 away, they will need to remove 3 more and then look to the tenth to see that if they remove 3 from that tenth, .07 (or 7 flats) would be left.
Discussion	After students solve each problem, ask questions such as, "What did you remove first?" "Why?" "Why did you need to trade the tenth?" "How would you know when you would need to trade or not?"

➤ Students should begin solving similar problems using base ten blocks and directly modeling problems the same way they did as a class. For additional support, blank 100 charts (10 x 10 cm) may be needed.

➤ When transitioning towards the numerical representation of subtraction with decimals, there generally is confusion around the value of a number if a zero is added to the right of the decimal number. For example, .7 will seem like a smaller amount than .70. When subtracting .7 - .34, for instance, students might either try to subtract the .7 from the .34, or not add a zero to the .7 and bring down the 4. Using the giant grid is a great opportunity to help students see that .7 and .70 is the same amount so they understand what they are actually subtracting.

➤ As students move toward adding numbers larger than one (for example, 2.45), students should be provided with multiple 10 x 10 cm grids so they can directly model such problems.

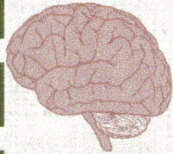

A Teacher's Insight

   It is important to have students draw in pictures what they are building with their tools as they are developing concepts around number and operation. This way they can eventually use their pictures as tools and wean themselves off of the more concrete models as they strive to internalize the meaning of a specific operation. Drawing reinforces the procedures and strengthens connections between representations.

   Make sure that students are literally drawing what they do with their tools. Once, I watched a 5th grade class solve a decimal problem with pictures and then with numbers. Each was so eloquently done, but one did not connect to the other, and when asked, they were unable to do so. In this case, students are not using a picture to derive or show how a procedure works, just as another way to solve the problem.

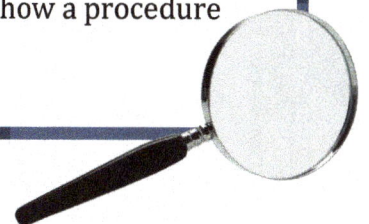

Building the Multiplication Chart

Grade Level: 3

Materials: Giant 100 grid that is blank, 100 sheets of paper that is cut into the size of the squares of the grid, or 100 base ten flats. Dry erase marker.

Objective: In the context of building arrays, students will understand the relationship of arrays to the multiplication chart.

*Note: The exploration portion of this lesson works for up to about 5 students. If working with a larger group, a small group can use the giant 100 grid while other groups work with a smaller grid and color tiles. The Introduction and discussion can be done with the entire group around the grid.

Common Core Connection:

Grade 3

CCSS.Math.Content.3.OA.A.1 Interpret products of whole numbers, e.g., interpret 5 × 7 as the total number of objects in 5 groups of 7 objects each. *For example, describe a context in which a total number of objects can be expressed as 5 × 7.*

CCSS.Math.Content.3.OA.B.5 Apply properties of operations as strategies to multiply and divide.[2] *Examples: If 6 × 4 = 24 is known, then 4 × 6 = 24 is also known. (Commutative property of multiplication.) 3 × 5 × 2 can be found by 3 × 5 = 15, then 15 × 2 = 30, or by 5 × 2 = 10, then 3 × 10 = 30. (Associative property of multiplication.) Knowing that 8 × 5 = 40 and 8 × 2 = 16, one can find 8 × 7 as 8 × (5 + 2) = (8 × 5) + (8 × 2) = 40 + 16 = 56. (Distributive property.)*

CCSS.Math.Content.3.OA.C.7 Fluently multiply and divide within 100, using strategies such as the relationship between multiplication and division (e.g., knowing that 8 × 5 = 40, one knows 40 ÷ 5 = 8) or properties of operations. By the end of Grade 3, know from memory all products of two one-digit numbers.

Introduction	Ask students what an array is and to make one using the blank 100 grid and the pieces of paper or base ten flats. They should be instructed to start in the top left corner when building their array. Ask them what number sentences can be written about this array (students may provide addition, multiplication or possibly division). Have students tell how many are in total and then remove the final piece of paper (lower right corner) and write that total with the dry erase marker on that square. An example is shown below:

			15						

Exploration	Inform the students that they will be building all the arrays that they possibly can on the 100 grid and recording the total tiles (papers) after completing each one. They will know that they have completed all the arrays when all of the squares have been filled up with numbers. What might happen is that students try to fill out the grid like a hundred chart without making all the arrays, but you need to encourage them, at this point, to keep building the arrays. They will soon realize that it will not work if they just make a 100 chart. After completing about a third to a half of the arrays, they will begin to see that they are skip counting, perhaps with the tens. If they start filling in the numbers by skip counting, ask them questions, such as why they are choosing to put that number, what patterns they are noticing, etc.

| Discussion | As a whole class, sitting around the giant 100 grid, once it is filled out, students should discuss what they notice. Many will notice some patterns, such as the tens. Ask them why there are so many of the same number, such as the 9 or 12. They might not know how to respond, so build one of those numbers and build another. You should guide them to notice that there are many ways to make the same number, and record them on the board (for example: 2 x 6, 3 x 4, 6 x 2, 4 x 3). Ask them what they notice. They should notice that there is a 2 and a 6 in two of the problems, etc. Have students build each one on the 100 chart and allow them to see that the arrays are simply rotated, from one to the other. Invite students to come up and find a duplicate number and decide if it is a result of a rotated array or if it is made from a different array. Introduce the term *commutative property* to describe these arrays and multiplication problems.

Ultimately, you want students to realize that what they have just created is a multiplication chart, something that is probably hanging somewhere in their classroom. At this point, this chart will have so much more meaning. |
|---|---|

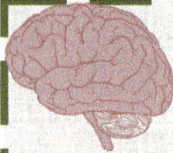

Did you know? #9

The The pre-frontal cortex is located on the frontal lobes of the brain. Elkhonen Goldberg refers to it as the "executive brain" which "gives us our interpersonal abilities and plain old common sense; for example, the ability to 'read' situations, discern the meanings of facial expressions, and anticipate the consequences of various actions," (Dennison, 2006, p. 57).

Make it Memorable

Help Students connect new material with movement to make retaining ad recalling the information easier by:

Hop, jump, leap, crawl, or hop like a frog to the different
Points of a geometric shape (possibly taped on the floor)

Perimeter

Grade Level: 3

Materials: Giant 100 Grid, post-it notes, dry erase marker, graph paper, handout

Objective: In the context of using the giant 100 grid, students will explore the idea of perimeter of a regular and irregular shape.

> ### Common Core Connection:
>
> #### Grade 3
> CCSS.Math.Content.3.MD.D.8 Solve real world and mathematical problems involving perimeters of polygons, including finding the perimeter given the side lengths, finding an unknown side length, and exhibiting rectangles with the same perimeter and different areas or with the same area and different perimeters.

Introduction **1** regular shapes	As students are gathered around the 100 grid mat, place a post-it at any 4 corners, creating a smaller rectangle. Ask a student to use a dry-erase marker to connect the 4 corners. Ask how many of the segments, or units, are under the marker – or are around the rectangle – and how they know. Label with numbers and it might even be necessary to show them how to put tic marks on each unit segment to keep track of how they are counting. Ask them to draw the shape and numbers onto a graph paper.
Exploration **1**	Divide the class into 4 groups and have each group at each corner of the mat. Allow one student from each group to create rectangles by placing post it notes on 4 intersecting points. Students record the size of the rectangle on graph paper and calculate the perimeter. Do this a few times so they have had enough practice.
Discussion **1**	Have a couple groups share one of their rectangles and how they solved for the perimeter. Ask if they are noticing a pattern and if they have a faster way to find the perimeter (if they are counting all). Some students will notice that the sides are the same and they only need to count one of the two and then double the number. Ask the class to help generate a number sentence and then a generalized equation (l+l+w+w; 2l+2w), for solving for perimeter. The children will most likely not be familiar with variables, so they need to be introduced naturally as a way to not have to use the entire word. They also will not know about placing a letter next to the variable, so an "x" can be used if necessary. Ask students what shape might have the equation of 4 x s (side).

Introduction 2	Review what was generated from the previous lesson and on the 100 grid mat create a new shape with the dry erase marker that is something like: 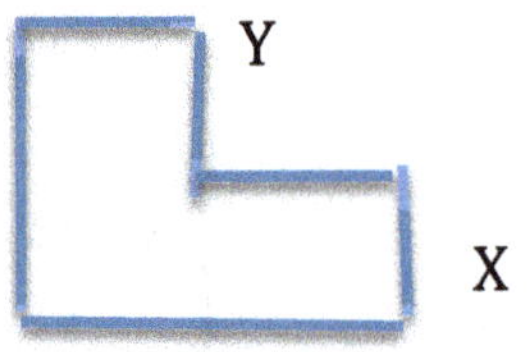
Irregular shapes	
Exploration 2	Have students calculate the perimeter, which they will do by counting. Ask them if they know how long side X is, for example, would there be a way they could solve for side Y. The trick here is to get students to see that the side across from X and Y is the same size as X and Y combined. Do the same with the other sides as well.
	Erase and create a new shape on the mat and ask similar questions. If students are starting to get the idea, split them into small groups and they can create the same shapes on each of 4 corners of the mat, which they record onto their graph paper. Have them note on their paper the sides that are equivalent (one longer and two shorter, such as in the illustration above).
	Provide the handout included for students to solve in partnerships within small groups – so the two sets of partners can be working together. This handout will be more abstract than what they have just been doing because there are no grid lines and they have to rely on the numbers presented. If it is too abstract, draw on grid paper or the 10 giant grid for support.
Discussion 2	Pick one of the problems that proved to be challenging and as a class discuss how they solved for perimeter, by first discussing the length of the missing sides.

Note: This activity may need to be done with older students who are having trouble with the concept of perimeter.

Find My Perimeter

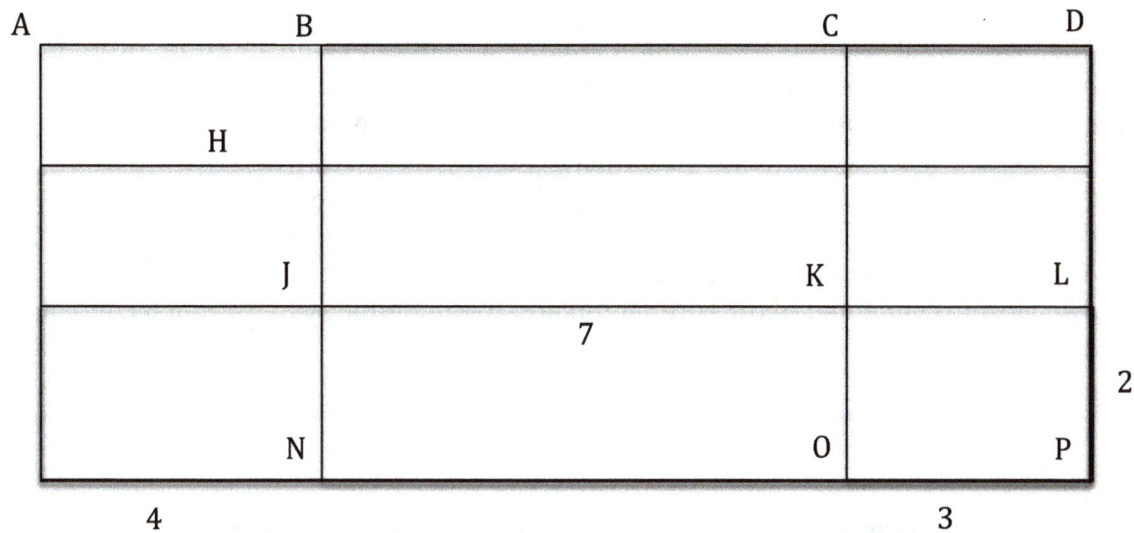

What is the perimeter of the following shapes?

ABFE_____   FGKJ_____   ABNM_____

KLPO_____   JLPN_____   BDPN_____

ABFGOM_____   JKCDPN_____

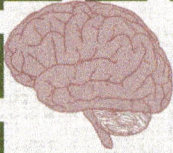

The Pre-frontal cortex is the decision making part of our brain and is involved in making plans, judgments, motivation, and impulse control. It "enables our conceptual and abstract thinking and our ability to reason and change our conscious concepts and ideas" (Blomberg & Dempsey p. 107), and is the part of the brain that is last to develop. It is also the part that is the most susceptible to damage in adolescents who engage in smoking marijuana. Like other parts of our brain, the pre-frontal cortex develops via our movement and sensory skills. It is also closely connected to the cerebellum and to the limbic system, which controls our emotions.

A Teacher's Insight

At my school, we have our playground equipment on black spongy square tiles. One section of the playground, for whatever reason, has green tiles of about an area of 360 tiles. It is in the following shape: This was perfect for students to count, draw on grid paper, and solve for perimeter and area. Lessons in this section are not exclusive to use with the 100 grid as opportunities may be naturally found naturally in your local school environment.

Coordinate Graphing

Grade Level: 5 / 6

Materials: Giant 100 Grid, post-it notes, dry erase marker

Objective: In the context of graphing points, students will see the relationships between them and generate a "rule" or equation for the line created.

Note: It is best that the class has data pre-collected. This might have been from earlier in the year, perhaps when gathering information about the class, or during a science experiment. It is also fun to gather data on rate, such as how fast to run a lap and then project it to 2, 3, and 4 laps, or perhaps on how many times a student can dribble a ball in one minute.

Common Core Connection:

Grade 5

CCSS.Math.Content.5.G.A.1 Use a pair of perpendicular number lines, called axes, to define a coordinate system, with the intersection of the lines (the origin) arranged to coincide with the 0 on each line and a given point in the plane located by using an ordered pair of numbers, called its coordinates. Understand that the first number indicates how far to travel from the origin in the direction of one axis, and the second number indicates how far to travel in the direction of the second axis, with the convention that the names of the two axes and the coordinates correspond (e.g., x-axis and x-coordinate, y-axis and y-coordinate).

CCSS.Math.Content.5.G.A.2 Represent real world and mathematical problems by graphing points in the first quadrant of the coordinate plane, and interpret coordinate values of points in the context of the situation.

Grade 6

CCSS.Math.Content.6.NS.C.8 Solve real-world and mathematical problems by graphing points in all four quadrants of the coordinate plane. Include use of coordinates and absolute value to find distances between points with the same first coordinate or the same second coordinate.

Introduction	➤ Using a coordinate grid on the board, or interactive white board, explain the term axis and that there are two called "X" and "Y". Come up with creative ways to notice which axis goes where, such as Y stands for "you" which stands up. Give a student with an index card or post-it note with the "X" written on it and ask him to place it on the giant grid and explain why he put it there. Ask another student to then place the "Y" in its proper location. Explain to students that each line represents a number in an equal interval, such as 1, 2, 3, or 2, 4, 6, etc. The intervals will be determined by the data collected. ➤ Start with a simple ratio, such as: for each glass of lemonade you need 2 lemons and 1 cup of water (2:1). Have the students decide which axis should represent the water and which the lemons. Using post-its, mark the grid (1-10) on each axis. ➤ Ask a student to place a different colored post-it on the location that would indicate 2 lemons and 1 cup of water. When complete, ask students how many lemons and how much water would be in two glasses of lemonade. Have them place the post-it where it should go. Do the same for 3 glasses, then have students look at the pattern and predict where the next one would go. ➤ What is the relationship between the locations of each post-it. Have students connect them with a straight line using a dry-erase marker. Have students write the ordered pairs that were created onto a T-chart in their notebooks, and one on the board. Discuss what is the "rule" or "equation" that describes the relationship between the amount of lemons and water (2x = y or 1/2x = y, depending on which item went on which axis).
Exploration	➤ Students continue the lesson with a partner using graph paper, and other data sets that were previously generated in the class that have a linear relationship. ➤ Remind students to choose appropriate intervals for their data sets so that they have enough room to plot at least 5 points. Have them create a T-chart with the ordered pairs, graph and generate the equation that matches the relationship of the particular data set.

Discussion	<ul><li>Have students share out, on a document camera if possible, the data set they chose, how they chose to set up and label their graph, and the equation that represents their data set. Ask questions such as what they chose for their interval and why.</li><li>How can they come up with the equation by looking at their graph? Select equations, such as $1/2x=y$ and $1/3x=y$ and have students compare them to their respective graphs. See if students can notice that the points go up one and over two, which match the $1/2x$. After connecting to some simple equations to the graphs, pose a more complicated graph where the distance between the points are up 2 and over 3. When students notice that it is up 2, over 3, and write it as $2/3x$, then have them see if they plug in X points from their ordered pairs if they will get the corresponding Y.</li><li>Announce that they have just discovered "slope", defined as rise/run.</li></ul>

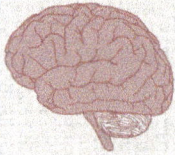

Did you know?

In the last couple of decades, mostly from the work of Michael Merzenich, neuroresearchers have begun to understand that the brain is not rigid, rather is plastic and capable of change until the day we die. In fact, back in 1949, Donald Hebb was the first to propose that learning linked neurons. He suggested that "when two neurons fire at the same time repeatedly, chemical changes occur in both, so that the two tend to connect more strongly," (Doidge, 2007, p. 63). Michael Merzenich, one of this country's most renoun neuroscientist today, expanded on this by saying that strong connections are made when they are activated at the same time.

He explains that when "we perform an activity that requires specific neurons to fire together, they release BDNF" (p. 80), (brain-derived neurotrophic factor) a growth factor which helps neurons to wire together so they fire together in the future. BDNF also helps with the mylenization of the neurons to speed up the impulses (Doidge, 2007).

Make it Memorable

Help Students connect new material with movement to make retaining ad recalling the information easier by:

Jump roping while reading addition facts posted on a wall

Experimental vs. Theoretical Probability

Grade Level: 7

Materials: Giant 100 grid (or number line), giant die

Objective: In the context of rolling a die and creating a human bar graph of outcomes, students will find the experimental probability of rolling a die and compare it to its theoretical probability.

Common Core Connection:

Grade 7

CCSS.Math.Content.7.SP.C.7 Develop a probability model and use it to find probabilities of events. Compare probabilities from a model to observed frequencies; if the agreement is not good, explain possible sources of the discrepancy.

CCSS.Math.Content.7.SP.C.7a Develop a uniform probability model by assigning equal probability to all outcomes, and use the model to determine probabilities of events. *For example, if a student is selected at random from a class, find the probability that Jane will be selected and the probability that a girl will be selected.*

Introduction	Teacher will ask students what they think the probability would be of rolling a given number on the die. After some private think time, students will discuss and justify their thoughts with their neighbor, and then with the whole class. Define this as the theoretical probability of the number.
Exploration	The teacher will ask if they think that it would always be true. As a way to investigate, one student will be asked to roll a die (a giant one is preferable). Students one by one will line up above the corresponding number, either placed at the base of the 100 grid or on a giant number line. When all of the students are a part of the graph, the teacher can get a photograph from an areal view. Rather than students standing, they can place an object in the corresponding square on the grid, such as a 100 flat.
Discussion	The photograph can be displayed using a computer and projector. A discussion should center on what the experimental probability for each number was and how it compared to the theoretical and why there might be a discrepancy between the two.

Next steps: Students can then explore theoretical and experimental probability for other objects such as 2 dice, two coins, one coin and one die, pulling color tiles out of a paper sack, etc.

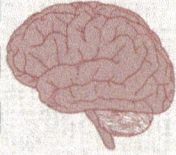

Did you know? #12

Merzenich worked a lot with brain maps and stumbled on the realization that the brain is dynamic and it could re-organize its maps depending what we do over the course of our lives (Doidge, 2007). Upon removing a monkey's finger, it was found that the part of the brain once in control of that finger was now taken up by something else. He explains that, "when we learn a bad habit, it takes over a brain map, and each time we repeat it, it claims more control of that map and prevents the use of that space for 'good' habits. That is why unlearning is often a lot harder than learning, and why early childhood education is so important—it's best to get it right early, before the 'bad habit' gets a competitive advantage," (Doidge, 2007, p. 60). Merzenich also found that as brain maps got bigger, the individual neurons get more efficient.

A Teacher's Insight

Probability lessons, once taught in younger grades, are postponed from the common core until grade 7. If your state has not adopted Common Core, this lesson may be appropriate, perhaps with some modifications, from grades 3-6.

Yardsticks, Rulers, & Tapes

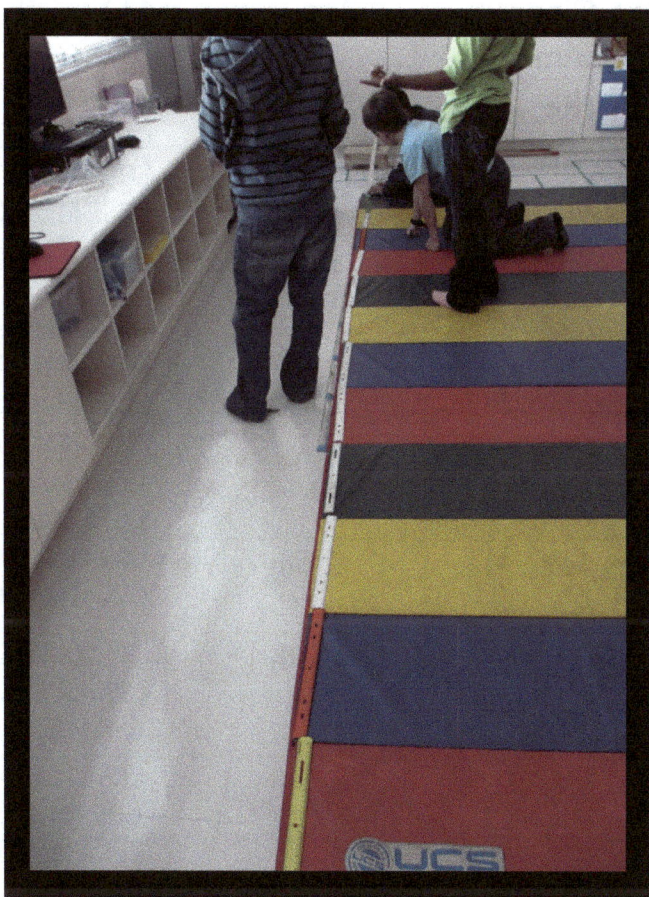

Lessons with Measurement Tools

Lesson Name	Grade Level	CCSSI Standard
Human Number Line	Kinder 1st	CCSS.Math.Content.K.CC.A.2 CCSS.Math.Content.1.OA.C.5
Non-standard Measures	1st & 2nd	CCSS.Math.Content.1.MD.A.1 CCSS.Math.Content.2.MD.A.2
How Long is 100 Feet?	Any	CCSS.Math.Content.2.MD.A.3 CCSS.Math.Content.4.MD.A.1
Perimeter: What Goes Around Comes Around	3rd	CCSS.Math.Content.3.MD.D.8
Area: Its all in the Middle	3rd	CCSS.Math.Content.3.MD.C.7 CCSS.Math.Content.3.MD.C.7a CCSS.Math.Content.3.MD.C.7b CCSS.Math.Content.3.MD.C.7c CCSS.Math.Content.3.MD.C.7d
Area of a Playground	3rd	CCSS.Math.Content.3.MD.C.7 CCSS.Math.Content.3.MD.C.7a CCSS.Math.Content.3.MD.C.7b CCSS.Math.Content.3.MD.C.7c CCSS.Math.Content.3.MD.C.7d
Measuring the School	4th	CCSS.Math.Content.4.MD.A.1 CCSS.Math.Content.4.MD.A.3
Collecting Data for Function Tables	4th & 5th	CCSS.MATH.CONTENT.4.OA.C.5 CCSS.MATH.CONTENT.5.OA.B.3
Throwing Contest	4th – 6th	CCSS.MATH.CONTENT.4.MD.B.4 CCSS.MATH.CONTENT.5.MD.B.2 CCSS.MATH.CONTENT.6.SP.B.4
Planning a School Relay	4th & 5th	CCSS.Math.Content.4.MD.A.2 CCSS.Math.Content.5.NF.A.2 CCSS.Math.Content.5.MD.A.1 CCSS.Math.Content.5.NF.B.4
Build Me a Bookshelf	6th	CCSS.Math.Content.6.RP.A.3 CCSS.Math.Content.6.G.A.4
Playground Equipment	6th	CCSS.Math.Content.6.G.A.4 CCSS.Math.Content.6.RP.A.3d
Circumference of Trees	4th & 6th	CCSS.MATH.CONTENT.7.G.B.4

Note: Grade level indicates where they appear in the Common Core, however, it may be appropriate to do a lesson in other grade levels depending on conceptual understanding.

Human Number Line

Grade Level: K/1

Materials: Construction paper, each with a large number written on it (one per student starting with "1") laminated (or commercially produced giant walk-on number line), large dice.

Objective: In the context of creating a human number line, students will practice the counting sequence and

Common Core Connection:

Kinder
CCSS.Math.Content.K.CC.A.2 Count forward beginning from a given number within the known sequence (instead of having to begin at 1).

1st grade
CCSS.Math.Content.1.OA.C.5 Relate counting to addition and subtraction (e.g., by counting on 2 to add 2).

Do this activity where there is space for all the students to stand in one line, preferably with blank wall space behind them. Can also be done in a hall.

Introduction	Raising a number written in the center of a piece of construction paper in random order from the pile for all to see, ask students what number it is, what number comes before and after. Do this with all the numbers and use this as a formative assessment.
Exploration	Pass out one number per student. Ask students to begin counting and as each number is said, the student holding that number is to raise it up. Do this again, this time, as each number is called, the student with that number is to go to the front and line up in the order of the numbers, displaying their numbers in front of them. If they are standing in front of a wall, have them turn around and tape them to the wall. If not, have them set them on the floor (all numbers facing in the appropriate direction) and tape them down.

	Allow the students to step away and to look at what they created. Many students should notice that they are the number line, or may even say that it is like a ruler. Now they have meaning behind a tool that perhaps they have been using. (phase 1- the next/previous number) Tell students that they are going to practice saying what number comes next. Roll a large die. A selected student should stand on the number rolled, for example "4," and then jump to the next number, shouting its name. Do this several times with different students and different numbers. You may want the entire class to shout with the jumping student. Do the same as above, but stating the previous number (-1) *Child can simply be standing on a number with eyes closed and asked to move forward (or back) one and tell which number she is standing on.* (phase 2 – counting on) Tell students that they are going to practice counting on. Roll a large die. A selected student should stand on the number rolled, for example six, and then walk up the number line, saying the numbers until he/she is stopped by the teacher, i.e., "7, 8, 9 10." Do this several times with different students and different numbers. Try using a number cube with larger numbers on it, or just pulling a number card from a basket. (phase 3 – addition, counting on) Do this when students are ready for simple addition. Roll a large die. If on, for example, "4", have 4 students stand on numbers 1, 2, 3, 4. Write "4 +" on the board. Then roll the die again and write the second addend. Have that many students stand up and as they stand on the subsequent numbers, the entire class should say those numbers. For example, if a 2 was rolled, then 2 students stand on 5, and 6. As #5 stands on his number, students say "5" and then #6 stands on his number and the students say "6".
Discussion	Without anyone walking on the number line, call out a number to the class and ask them to count up from that number. Ask them how they knew from where to start. The most common error at this age is to start with the same counting number and not the next one. Modify the discussion to fit the appropriate phase above.

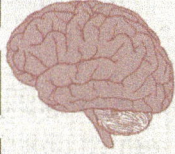

The brain does not like change. Therefore, in order for changes to take place in the structure of the brain, we need to have access through all the senses, including the proprioceptive receptors of the muscles. In addition, the brain needs to be engaged, there needs to be repetition, or rehearsal of activity, and feedback to the brain is necessary. It is said that it takes exactly 3 weeks, 21 days, for connections to be made in the brain (Gold, 2008), so consistent repetition is necessary until the appropriate connections are made.

Make it Memorable

Begin each session, such as morning work or after lunch, with a run or exercise that reduces stress and increases communication between both hemispheres of the brain. Examples are deep breathing and cross-lateral activities, like cross crawls.

Non-standard Measures

Grade Level: 1ˢᵗ and 2ⁿᵈ

Materials: Multiple objects, such as color tiles, markers, unsharpened pencils (so that they are the same size), etc., recording sheet

Objective: In the context of measuring large objects from the environment, students will develop a sense of what measurement means and develop their own measurement tool as well as articulate that the smaller the unit, the more of them there will be.

Common Core Connection:

First grade
CCSS.Math.Content.1.MD.A.1
Order three objects by length; compare the lengths of two objects indirectly by using a third object.
CCSS.Math.Content.1.MD.A.2
Express the length of an object as a whole number of length units, by laying multiple copies of a shorter object (the length unit) end to end; understand that the length measurement of an object is the number of same-size length units that span it with no gaps or overlaps

Second Grade:
CCSS.Math.Content.2.MD.A.2
Measure the length of an object twice, using length units of different lengths for the two measurements; describe how the two measurements relate to the size of the unit chosen.

Introduction	Select an object from the room and ask the students how big is it. Students are sure to respond with a variety of different answers, some of which may involve comparing it to something else. Ask them what they should use to compare it too. Select one of the objects they have chosen and ask how they should use it to compare. Invite a couple of student to try to compare with the chosen object, as they typically will not put them end to end and thus end up with a different answer. If so, ask the class why when measuring with the same object they might get a different answer. There may be multiple factors, such as not starting at the end of the item being measured or putting too much space between the measuring tool. Discuss how it is important that both objects have the same starting point and that when measuring the objects need to be touching.

Exploration	Divide the students into partnerships and tell them that they will measure, or compare their partner to something else. In their math journals, or on a recording sheet they are to write what they used to measure their partner and what object they chose. Encourage to try to use at least two different measurement tools when measuring.

Object Used to measure partner	How many did it take?

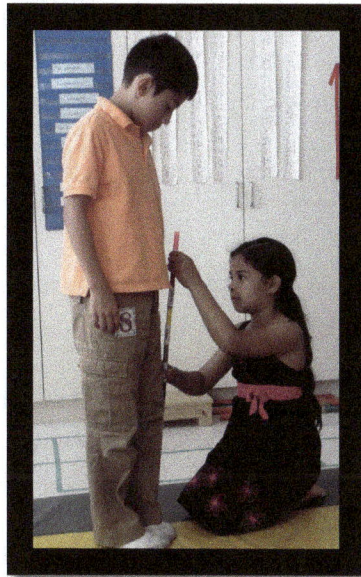

Discussion	Bring the class together and have them share what they used and how many of each tool. After several share outs, start a discussion about some misconceptions that you noticed while the students were measuring. There will still be problems regarding gaps and starting at the end. Choose a student who used two different objects that were significantly different in size, such as a marker and a color tile. Ask the class which would there be more of when measuring their partner. They may say the marker because it is larger, therefore they think there would be more of them. But if asked to reflect back on the data shared, they will notice that there were more of the smaller item. A big idea they need to encounter is that the smaller the measuring tool, the more of them it will take.

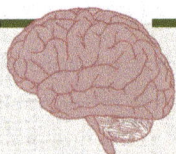

A Teacher's Insight

The big idea in this lesson, the smaller the measuring tool, the more of them there will be compared to a larger tool, is a very important concept. This big idea links to fractions when students have to rationalize why ½ is larger than ¼ even though the "4" is a bigger number.

How Long is 100 feet?

Grade Level: Any

Materials: 100 foot measuring tape

Objective: In the context of walking on the 100 foot measuring tape, students will internalize a sense of how long 10 feet is.

> ### Common Core Connection:
>
> *Grade 2*
> CCSS.Math.Content.2.MD.A.3 Estimate lengths using units of inches, feet, centimeters, and meters.
>
> *Grade 4*
> CCSS.Math.Content.4.MD.A.1 Know relative sizes of measurement units within one system of units including km, m, cm; kg, g; lb, oz.; l, ml; hr, min, sec. Within a single system of measurement, express measurements in a larger unit in terms of a smaller unit.

Introduction	Outside on the playground, ask the students how far they think 100 feet is from where they are standing. Notice the differences in responses. Have a couple of students take the measuring tape and pull it out as long as it will go and lay it down flat, foot side up. Discuss discrepancies with their estimations.
Exploration	Students break off into partnerships (if there is an odd number, then the teacher can be a partner) and make a line behind the zero. Instruct them to walk for 10 feet and stop. Then walk ten feet again, stop. Repeat until they reach 30 feet. Now have one of the two students close his eyes and walk 10 feet and stop where they feel that 10 feet should be. Stop and open the eyes to verify. Readjust by starting on the next 10. Continue until they reach the end. They can start over so the other student can close his eyes. The teacher will need to space the partnerships out so they are not walking on top of each other.
Discussion	Discuss what they noticed when walking. Many students will be surprised that they were able to do it accurately after about the third try.

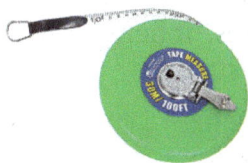

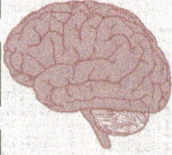

Moving our muscles produces proteins, like IGF-1 and VEGF, which travel to the brain through the bloodstream and affects the pre-frontal cortex allowing children, for example, to have the ability to stop and consider a response before making a decision. Exercise also balances neurotransmitters, and research has shown that simply jogging thirty minutes two –three times a week improves executive function (Ratey, 2008). In addition, learning complex movement sequence stimulates the pre-frontal cortex and improves learning and problem solving.

Make it Memorable

The Brain Dance, a video by Ann Green Gilbert, is a great way to get students to engage in many necessary movements in a quick 6 minutes. It helps prepare their bodies for learning and can aid in transitions, such as returning from lunch.

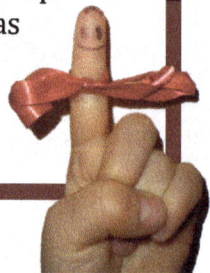

Perimeter:
What Goes Around Comes Around

Grade Level: 3rd grade

Materials: Variety of measuring materials, including standard and non-standard units (such as rulers, measuring tape, 1" tiles, cubes, markers, or anything that can re-iterate to compare to another object), 1" or 1cm square paper, electrical tape, section of floor made with square tiles

---

Common Core Connection:

3rd Grade
CCSS.Math.Content.3.MD.D.8 Solve real world and mathematical problems involving perimeters of polygons, including finding the perimeter given the side lengths, finding an unknown side length, and exhibiting rectangles with the same perimeter and different areas or with the same area and different perimeters.

Objective: In the context of measuring different objects in and/or out of the room, students will understand that the term "perimeter" refers to the distance around an object, as well as generate strategies to figure out what the perimeter is of that object. Students will begin to generate informal language that relates to the formula for finding perimeter of rectangular objects.

Note: This lesson will take the course of multiple days. Length will depend on the readiness of the students.

Introduction 1	Using electrical tape or painter's tape, mark off a section of 1' floor tiles (either in the room or perhaps a hallway), perhaps 3' x 4'. If possible make a few of these to be used by multiple groups. Each rectangle could be a different dimension so that groups could rotate, if desired. Mark off these areas while students are watching, if possible so they can see the roll of tape getting smaller and smaller.
	After you mark off a rectangle on the floor. Ask students what they can do to figure out how much of the tape was used. Students will probably naturally just start counting tiles. If they do not know what to do, just ask them to use the tiles to help them.

Students tend to count the tiles on the inside of the perimeter as shown below:

1	2	3	4
10			5
9	8	7	6

By doing so, they are not counting the two sides of the corner squares. To have them prove if they are correct or not, cut a straw or string that is the size of the tile and lay it on the tape from one of the corners. Ask them if they think if it will take ten of them to go around. Have them partner talk to decide, then share out as a whole group.

Have students cut several pieces of what you used to demonstrate and try it out. When they realize it takes 14 pieces, ask why it is not 10 pieces and what did they do, or not do, when counting tiles. Another thing that should be brought up is how they could know how long one side is by knowing the length of another. This will most likely come up as students are laying down the string and when one side is full, they will automatically state how many would fill the side directly across it.

Exploration 1	Explain the difference between standard and non-standard units. Explain that if they used a piece of string, for example, and wanted to explain to someone on the phone how large the rectangle was, they could not merely say that it was 14 pieces of string or the other person would not be able to visualize it. Ask students how long a ruler is. They should all be very close to the actual size of a ruler. Tell them that they are going to use rulers to figure out how many of them it will take to go around.

	As students start measuring their assigned rectangle, if the tiles are 1 square foot tiles, they should immediately notice that they are the same size. This will help make the 1-foot tile as a benchmark to use in future measurements. If perhaps the tiles are 6", then they will see that one ruler fits over two tiles. On the grid paper, have students draw the particular rectangle that they are measuring and label with the respective lengths. Each square should represent one ruler length.
Discussion 1	Share out drawings from the different groups because some may not know how to appropriately label the lengths.
Exploration 2	Give the students the opportunity to measure large objects in the room, or in other parts of the school, such as the playground, cafeteria, or gymnasium with rulers. More than likely there will not be enough rulers to measure the entire perimeter, which will cause the students to have to figure out what to do when they run out of rulers. Something to watch out for is if students are putting gaps between the rulers.

Discussion 2	When finished, students need to figure out how many ruler lengths are around the object. At this point, they should be starting to articulate two lengths and two widths when explaining the perimeter of rectangular objects. Since different groups are measuring different objects, help them generalize this by writing equations such as 2 lengths + 2 widths = perimeter of object; $2l + 2w = p$.
Exploration 3	Pick one of the objects and ask students what would happen if they measured it in inches. Would they have a larger or smaller answer. Surprisingly, many students will say a smaller answer. They say this because the unit is smaller and they do not realize tha a smaller unit yields more of that unit, thus a larger answer.

Intiiate a discussion about an inch and how a square color tile is one square inch and show them that 12 of them fit on the ruler. You can even tape 12 of them together to make a ruler. The point to emphasize is that one inch on the ruler is the point where the first one inch square tile ends.

Have student groups return to the object that they measured and have them figure out how many inches this will be. Interestingly, students will actually re-measure their object with the rulers, even though they know how many rulers. This is because they are now focused on the number of inches rather than the number of rulers. |
| Discussion 3 | Since their object will be quite large, the will be required to add many 12's. This discussion will most likely focus on strategies that they used to add all of their 12's. This can very well lead into strateies of multiplying tens and ones, thus giving a hands-on experience for the distributive property, even if it has not been formally introduced.

Have the class confirm whether or not their answer was larger or smaller than when they measured with rulers (feet), and why it would make sense for that to be true. Also, have them test if the equation that was posed during the Discussion 2 held up even when measuring with inches rather than feet. |

A Teacher's Insight

Students often get the terms *area* and *perimeter* mixed up. My theory has been that they are taught too closely to one another and not enough exploration of either one to give them justice. Therefore, students have not internalized the differences. One way to prevent this is to:

1) Heavily reinforce the term *area* when teaching multiplication with arrays and the area model.
2) When addressing students to sit on the class carpet, invite them to either sit on the *perimeter* when you want a circle or the *area* when you want them bunched up close.
3) When conducting explorations of area and/or perimeter, have students use the target vocabulary a lot when discussing. It has been said that we need to use a work appropriately about 9 times before we own it.

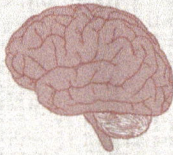

Did You know?   #16

Dr. Harold Blomberg, a Swedish psychiatrist, explains:

"The frontal lobes of the cortex receive important stimulation from the cerebellum with connections going to both the prefrontal cortex and the Broca speech area in the left hemisphere. In cases of dysfunctions of the cerebellum, these areas may not develop properly causing problems with speech development or difficulties with attention and the ability to make judgments, control impulses, motivation, and making sustained effort. The basal ganglia also have important nerve pathways to the prefrontal cortex. Therefore, when these two areas are not well linked know that there are ...motor problems that will then be part of the ba problems with attention and impulse control" (108). "If the nerve n between the prefrontal cortex and the limbic [emotional] sys are not sufficiently developed, or if the prefrontal cortex is not ciently stimulated from the cerebellum or the basal ganglia, we run a greater risk of switching off the prefrontal cortex and becoming overwhelmed by our emotions, causing fits of anger or anxiety (p. 111)."

Area of playground

Grade Level: 3

Materials: Something that has a large area made by square units that is not rectangular. Graph paper, rulers, clip-boards, pencils.

Objective: In the context of measuring an object with a large area comprised of an irregular shape students will understand that the area can be figured by breaking the shape down into smaller rectangular shapes and then adding them together.

Note: This lesson is challenging because it requires visual –spatial skills for counting the tiles and transferring the footprint to the graph paper.

Common Core Connection:

Grade 3

CCSS.Math.Content.3.MD.C.7 Relate area to the operations of multiplication and addition.

CCSS.Math.Content.3.MD.C.7a Find the area of a rectangle with whole-number side lengths by tiling it, and show that the area is the same as would be found by multiplying the side lengths.

CCSS.Math.Content.3.MD.C.7b Multiply side lengths to find areas of rectangles with whole-number side lengths in the context of solving real world and mathematical problems, and represent whole-number products as rectangular areas in mathematical reasoning.

CCSS.Math.Content.3.MD.C.7c Use tiling to show in a concrete case that the area of a rectangle with whole-number side lengths a and $b + c$ is the sum of $a \times b$ and $a \times c$. Use area models to represent the distributive property in mathematical reasoning.

CCSS.Math.Content.3.MD.C.7d Recognize area as additive. Find areas of rectilinear figures by decomposing them into non-overlapping rectangles and adding the areas of the non-overlapping parts, applying this technique to solve real world problems.

| Introduction | Find a large floor area, either indoors or outdoors, that is not fully rectangular. An outdoor playground made with square spongy material is ideal. If nothing is available, mark off a large portion of the cafeteria or other large room that is composed of square tiles. Make sure to make the section irregular and not rectangular.

Provide students with their materials and tell them that they need to find out how many tiles make up the designated section without counting all the tiles. Send students to work in groups of 2 or 3. |
|---|---|

Exploration	In their small groups, students will create a footprint on their graph paper and then figure out the number of tiles, without counting all.
Discussion	Have the different groups compare their pictures. If different groups have drawings with different dimensions, then they first have to prove which footprint is accurate. If most groups have the same drawing, then the consensus could be made based on that. However if there are just a few groups and most have different drawings from each other, make them go back to re-measure, and subsequently re-calculate. Do not discuss the calculations until there is consensus with the drawings. Once the class is ready to begin discussing the calculations, discuss how they figured out the area. If the sections they had to break apart were small, they may have used multiplication, but more than likely they used repeated addition for the different sections. At the end, highlight a team that used a multiplication number sentence to model the total area, for example: $(8 \times 7) + (14 \times 9) = 182$

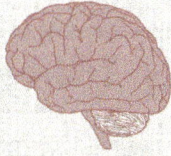

Did You know?   #17

For a nerve cell to grow, it receives stimulation from the senses and begins to myelinate, or to create a fatty coating, to make transmissions quicker. Movement causes the continued myelinization, growth of dendrites and axon terminals. When we move, chemicals are produced in the muscle, which results in new dendrites being sprouted. Therefore, repeated movements help to strengthen the neural pathways that run between the brain and the body (Goddard Blythe, 2007).

Make it Memorable

MeMoves is a DVD that contains a series of short, yet precise activities where students copy hand motions that are set to specific music that calms and relaxes the nervous system, helps focus the child, and improves bilateral integration of the brain. The easy implementation of the DVD makes it perfect for transitions in the classroom.

See http://thinkingmoves.com/ for more information.

Measuring the School

Grade Level: 4

Materials: 100 foot measuring tape and/or trundle; graph paper with small squares, clipboard, marker

Objective: In the context of measuring the buildings of the school, students will be able to estimate lengths in meters (walls) as well as be able to accurately use a measuring tape and round to the nearest meter. Students will also be able to find area and perimeters of irregular figures.

Common Core Standard(s) Addressed:
Grade 4
CCSS.Math.Content.4.MD.A.1 Know relative sizes of measurement units within one system of units including km, m, cm; kg, g; lb, oz.; l, ml; hr, min, sec. Within a single system of measurement, express measurements in a larger unit in terms of a smaller unit.

CCSS.Math.Content.4.MD.A.3 Apply the area and perimeter formulas for rectangles in real world and mathematical problems. *For example, find the width of a rectangular room given the area of the flooring and the length, by viewing the area formula as a multiplication equation with an unknown factor.*

Introduction 1 Measuring	Inform the students that they will be measuring the school with the purpose of creating a map. Possibly show them a map of a local area with several buildings, such as an office park. If they are unfamiliar with the 100ft measuring tape and/or trundle wheel, introduce the tools to them and explain how they work. The trundle wheel is the fastest tool, but using the measuring tape is more valuable. In small groups, I have 3 students using the measuring tape and two students using the trundle wheel, to verify the other tool's results. I have an additional student or two recording results on graph paper (one line segment per meter). After each wall or two, rotate students so they get even experiences doing all jobs.
Exploration 1	Before measuring any given wall, encourage them to estimate the length in meters. Have them measure to confirm. If the building is longer than the 100 feet, students will have to figure out what to do to determine the entire length. After each side is measured, the recorder has to draw it on the graph paper. This is considerably challenging for many students. First they will have to determine where on the paper to start drawing. Then, as each corner is turned, they have to figure out which way that is on the paper. The measurers will need to learn how to round to the nearest meter (I found that this is a perfect founding in context exercise, which strengthens this skill).

Discussion 1	Throughout the activity, the adult in charge of the group needs to be continually guiding and coaching the students along. After each building is measured, the different groups can come together to compare their measures and drawings. Any discrepancies need to be re-measured to ensure accuracy. Precision is key, especially when measuring irregular buildings, otherwise the start and end-points will not match
Introduction 2 Creating the map	Once all of the buildings have been measured and recorded, it is time to clean up the work and create the map that will be displayed in the school. Inform students that they will be re-drawing their pictures, but they must be done accurately and without random lines in the picture (this includes dots that many put in each square for counting).
Exploration 2	First, between all the groups, the buildings are re-drawn (unless each group will create their own map). The issue now is how to position the buildings. Students need to go and measure distances between buildings (the trundle wheel is best for this job) so they can properly position them.
Discussion 2	As the map is being positioned, allow students to justify where they are placing the buildings based on their measurements. If desired, students can mark classrooms, office space and bathrooms. If they are going to divide a building into classrooms, challenge them to measure where one begins and ends so they can be more accurate. Laminate map when complete.
Extension Determining the area & Perimeter	The school map can be the springboard for many other learning opportunities, such as determining the area and perimeter of each building, especially if there is a mix of regular and irregular shaped buildings.

School map

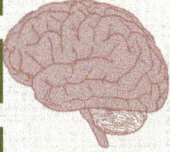

Did you know? #18

Stephen Lisberger at UC Berkeley found that to get a cell to fire in the cerebellum, the head must move at the same time as the target. For example, when a baby is creeping on hands and knees, the head continues to move looking at the hand that is placed forward, which is the target. "It is during this creeping time that the cerebellum becomes myelinated," (Gold, 2008, p. 142). In addition, when babies start to do repetitive rhythmic movements, there is rapid development due to the stimulation of the cerebellum. Children who are unable to rhythmically rock, like sliding up and down on his back with knees bent, may have a dysfunction of the cerebellum, (Blomberg & Dempsey, 2011).

A Teacher's Insight

This lesson is a favorite, and students look forward to this every year. The measuring portion alone of this project can take several days, depending on the size of the school. Our school consists of 10 buildings, some rectangular and some irregular, so we begin with measuring all the rectangular buildings and progressed to more complicated ones. If doing this with an entire class, split them up into groups of 6, where either one group is measuring while the rest of the students is doing classwork, or have each group measuring a different section or building, each group being monitored by an adult.

Collecting Data for Function Tables

Grade Level: 4-5

Materials: Clock or stopwatch with second hand, equipment for physical activity exercises such as jump ropes, hula-hoops,

Objective: In the context of engaging in physical activities, students will collect data for a function table and extend that to creating a line graph using the ordered pairs generated by the data.

Common Core Connection:

Grade 4
CCSS.MATH.CONTENT.4.OA.C.5
Generate a number or shape pattern that follows a given rule.

Grade 5
CCSS.MATH.CONTENT.5.OA.B.3
Generate two numerical patterns using two given rules. Identify apparent relationships between corresponding terms. Form ordered pairs consisting of corresponding terms from the two patterns, and graph the ordered pairs on a coordinate plane.

Introduction	Have students take their pulse for 10 seconds and then determine what their pulse would be in one minute. Discuss how they know and what they did to solve the problem. On an in and out box (T-chart, or function table), record the number of beats per 10 seconds of one of the students. Ask how many there would be in 20 seconds, 30, seconds, and so on up to 60 seconds. Determine what the rule would be for determining pulse for that individual.

Rule: $(S \div 10) \bullet 9 = P$

Seconds	Pulse
10	9
20	18
30	27
40	36

Exploration	Take students outside and have them engage in different physical activities where they can time themselves, via a partner. Example of activities are: ✓ Running a lap ✓ Jumping rope in 30 seconds ✓ Hula hoop ✓ Number of pull ups in 10 seconds ✓ Number of push ups in 10 seconds ✓ Number of sit-ups in 10 seconds After they collect the data from at least 4 different activities, have them graph the function on grid paper and determine the rule for the function. Allow students to illustrate the graphs so they can be posted in the room (make sure the function tables accompany the graphs).
Discussion	Share out the misconceptions that you noticed as students were collecting and recording data as well as generating the rule. When sharing out the different graphs, students should notice that their graphs are a straight line, or a linear function.

Note: The following recording form has the function table arranged vertically and horizontally. Many times students only see the table written in one way and cannot generalize its use when written differently. Having students record and analyze data using both horizontal and vertical tables helps make connections between the two.

In X	Out Y

In X				
Out Y				

Rule:

In X	Out Y

In X				
Out Y				

Rule:

86

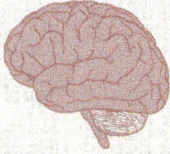

Did you know? #19

Research has shown that two areas of the brain that were associated solely with control of muscle movement, the basal ganglia and the cerebellum, are also important in coordinating thought. All learning, including that requiring abstract thought, occurs through movement, since abstract thought involves the internal repositioning of ideas. "Movement is the primary way that we integrate our learning into expressive action," (Dennison, 2006, p. 181).

Make it Memorable

Limit the presentation of new information, especially if cognitively demanding, to no more than 10 minutes.

Throwing Contest

Grade Level: 4-5

Materials: Football, kickball, Frisbee, or other throwing object, several pieces of sidewalk chalk, 100-foot measuring tape, clipboard, and paper.

Objective: In the context of having a throwing contest, students will practice measuring distances, record the class results on a graph, calculate, and analyzing the data.

Note: This lesson can be modified to include kicking or jumping.

Common Core Connection:

Grade 4

CCSS.MATH.CONTENT.4.MD.B.4

Make a line plot to display a data set of measurements in fractions of a unit (1/2, 1/4, 1/8). Solve problems involving addition and subtraction of fractions by using information presented in line plots.

Grade 5

CCSS.MATH.CONTENT.5.MD.B.2

Make a line plot to display a data set of measurements in fractions of a unit (1/2, 1/4, 1/8). Use operations on fractions for this grade to solve

Grade 6

CCSS.MATH.CONTENT.6.SP.B.4

Display numerical data in plots on a number line, including dot plots, histograms, and box plots.

CCSS.MATH.CONTENT.6.SP.B.5

Summarize numerical data sets in relation to their context problems involving information presented in line plots.

Introduction	Introduce to the class that they are going to have a throwing contest, but they are going to have to measure the distances of the throws. Divide the class into two teams. One team will throw while the other team has to measure.
Exploration	Students mark with chalk the starting point for the throw, kick or jump. The group that will measure need to have a keen eye to where the ball, or object lands so that they can run and mark the spot. Put some rules into place about who gets to mark, so that the entire group doesn't run to the spot all at once. By each mark, write the name of the thrower so they can later collect data. After each throw, allow him and another student to measure the distance of where the ball landed. A third student can record the result on paper. After the data is collected for the entire class, the list should be copied, projected, or otherwise visible to the class. In groups of 2 or 3, students will take the data and present as a graph. Students will need to decide which type of graph would be most appropriate and what interval should be used.
Discussion	The graphs should be displayed and discuss which shows the results most clearly. This graph should be used to learn concepts around data and statistics, such as mean, median, mode. Other problem solving questions, such as finding the difference between two scores, who scored better, the girls or boys, or What is the combined distance of the top 5 throws.

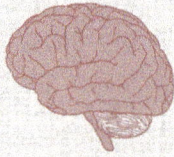

Movement does not necessarily mean that children have to be doing cartwheels during class, it may simply take form of talking, writing, knitting, or chewing, since different muscles of our bodies are being activated.   Every movement of the legs, arms, eyes, etc., result in some sensation going to the brain (Kokot, 2010). It "feeds information to the brain, helping to develop a sense of body map, of spatial awareness and body schema in relation to the self and to the environment." (Goddard, 2005, p. 47).

A Teacher's Insight

Anytime you plan to use data in a lesson, it is much more meaningful when the data has been generated by the students. They will be able to analyze it and create questions based off the data more easily than random data produced from a textbook. Since the numbers have meaning, the rules or equations that are generated take on new life.

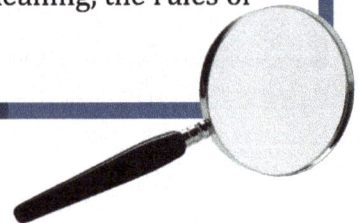

Planning a School Relay

Grade Level: 4th & 5th

Materials: Trundle Wheel, recording tools, sidewalk chalk

Objective: In the context of measuring the playground, or other open area, students will determine how many laps equal a mile. They will also plan out a relay ran with 4, 5 or 6 students on a team.

Note: Best done with a small group of students or partnerships.

Common Core Connection:

Grade 4
CCSS.Math.Content.4.MD.A.2 Use the four operations to solve word problems involving distances,

Grade 5
CCSS.Math.Content.5.NF.A.2 Solve word problems involving addition and subtraction of fractions referring to the same whole, including cases of unlike denominators
CCSS.Math.Content.5.MD.A.1 Convert among different-sized standard measurement units within a given measurement system (e.g., convert 5 cm to 0.05 m), and use these conversions in solving multi-step, real world problems.
CCSS.Math.Content.5.NF.B.4 Apply and extend previous understandings of multiplication to multiply a fraction or whole number by a fraction.

Introduction 1	Invite the students to help plan some end-of-the-year events for your school. They will first plan a one-mile race and then a relay. If they are unfamiliar with the trundle wheel, introduce it as a measuring tool.
Exploration 1	*Mile race* 1) With the trundle wheel, measure how many meters in one lap. Advise students to stay in the middle of the running space since measuring the edges will produce a deflated or inflated result. 2) Students figure out online, or from a different source, how many meters are in a mile (if the trundle wheel measures meters, or feet in a mile if the tool measures feet). 3) Now they need to divide the total number of meters in a mile by the number of meters in one lap. This will tell them how many whole laps in a mile and the amount left over (in tenths). 4) Now they have to figure out how many meters that amount of left overs represent. For example, the result is 7.6 laps per mile. If there were 210 meters per lap, then .6 of 210 (or .6 x 210; 6/10 x 210) = 126 meters. 5) Students need to go back and mark where 126 meters lands, thus the finish line.

Discussion 1	There is a lot of math involved in this activity, some which may be a review and others may be new, for example the idea of multiplication of a fraction/decimal by a whole number. Judgment will need to be used to decide which aspects of the activity need further discussion. *Depending on the level of the students, rather than dividing the total number of meters in a mile by the amount in each lap, perhaps they will need to create a two column table, consecutively adding the number of meters per lap until they reach the amount in a mile. Students can judge whether to round the left over up or down to the nearest lap.*
Exploration 2 Relay	*Relay* Using the data collected from the first activity, students have to plan out where students would stand if they wanted to have a relay with teams made up of different amounts of students. They would then have to judge which team make up would be the best choice and why. When doing this, students will have to divide the number of meters per lap by the number of potential runners to determine the number of meter each student would run. They should make a two-column table to record their results. It would be best if the group go out and test the different possibilities.
Discussion 2	As a class, after comparing and confirming the mathematics, students debate which relay situation would be optimal for a race among the entire grade level.

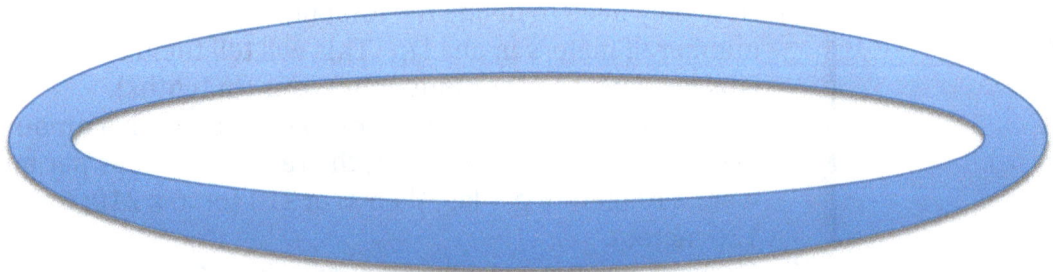

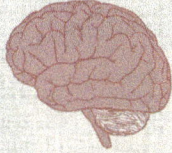

Learning happens when connections are made between nerve cells. When the body is inactive for 20 minutes or longer, there is a decline in neural communication (Kinoshita, 1997). This is why spending hours sitting in front of the television or playing video games is so harmful, especially for children with growing brains. In the classroom, this means that lessons that require students to sit and attend for long periods of time can be counterproductive.

Make it Memorable

Break a traditional 1-hour math lesson into smaller parts, such as:

- ✓ 5-10 minute introduction
- ✓ Private think time
- ✓ Partner talk about that which they reflected
- ✓ Whole group discussion
- ✓ Small group and individual work that allows students to explore a concept in more depth
- ✓ Small group share-out
- ✓ Whole group share out to help students make connections

Build Me a Bookshelf

Grade Level: 6

Materials: Cardstock, pencils, tape, yardsticks, rulers, scissors, graph paper (if needed).

Objective: In the context of creating a bookshelf, students will be able to appropriately use scale and calculate surface area of their bookshelf.

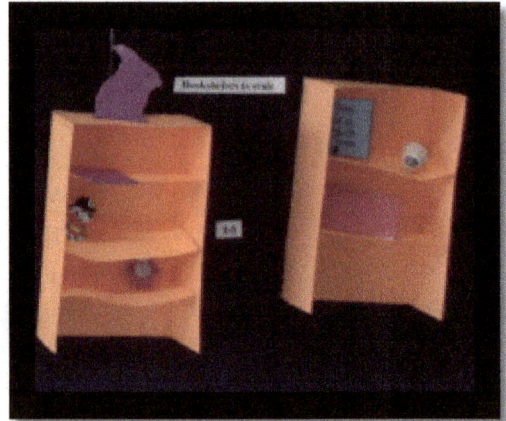

Note: This is a multiple day project.

Common Core Connection:

Grade 6
CCSS.Math.Content.6.RP.A.3 Use ratio and rate reasoning to solve real-world and mathematical problems, e.g., by reasoning about tables of equivalent ratios, tape diagrams, double number line diagrams, or equations.

CCSS.Math.Content.6.G.A.4 Represent three-dimensional figures using nets made up of rectangles and triangles, and use the nets to find the surface area of these figures. Apply these techniques in the context of solving real-world and mathematical problems.

Introduction	Ask students if they have ever been in a store and seen on display either a small tent or office furniture that looks like it was meant for dolls. Explain that the model is an exact replica of what they were purchasing, but proportionally smaller. When building a model, every dimension is divided by the same amount, which creates a scale or ratio. If one of the dimensions is off, then the entire model becomes disproportionate. This may be illustrated by using a picture on the computer. To proportionately enlarge it each side must be multiplied by the same amount. If one side is multiplied by a larger number the picture is no longer proportional.

	Find a simple bookshelf in the room and tell the students that they will be creating a model of it. Inform them that the model will be approximately a foot or so high, so they realize that they are not making it life-size. Ask for suggestions on how they might accomplish this. Students should realize that they need to divide the dimensions, but will not know by how much.

Start with one of the dimensions, like the height. Have a student measure it and ask them how tall it is, for instance 180cm. Have the group divide it by different amounts and figure out how tall it would be if they chose x amount. Allow each group to judge how tall they want their model to be and which ratio would work for them. For example, if they chose to divide by 10, their model would be 18 cm high and may be much too small, so dividing by 5 or 6 might be more desirable. If they chose 5, for example, then they need to maintain the 1:5 ratio throughout the entire measurement process. |
| **Exploration** | This project is best done with two students in a group. If having too many groups is too much with one bookshelf, which it probably will be, then do this with a few partnerships at a time, or if multiple bookshelves happen to be available, then use more than one.

Provide students the materials and have them begin their design process. This is going to be very tricky for some and you will have to guide the process by watching their errors. For examples, some students will measure all the dimensions and just record the results without drawing any kind of sketch. Keeping track of what the measurements represent is an important part of the process. |
| **Discussion** | As the projects are wrapping up, meet with the different partnerships and discuss what was their scale, check up on their calculations, ask about how they knew where to position the shelves and perhaps give tips on how to more securely attach the shelves. As a group students can discuss the parts that were the most difficult and what they learned about ratio and scale through this process. At the end, students may want to decorate their shelves with tiny home-made books and objects and the shelves can be displayed on a bulletin board. |

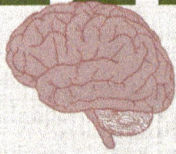

Neurological development, which partially takes place during the crawling (on belly) and creeping (hands and knees) process, determines the ability to read and write (Gold, 2008). Carl Delacado was a lead researcher who traveled around the world and found that societies that did not have a written language of its own were the same ones who did not permit their children to crawl or creep on the floor for whatever reason, such as in locations where crawling would be dangerous to the child.

A Teacher's Insight

For the students with poor visual-spatial skills, this entire project will be quite challenging, but is a necessary one. It may be wise to carefully pair students so that there is one who will be able to support the other visual-spatially. I do this lesson with my struggling learners, all who have struggles, so I need to carefully supporting all of them, at least in the beginning.

Playground Equipment

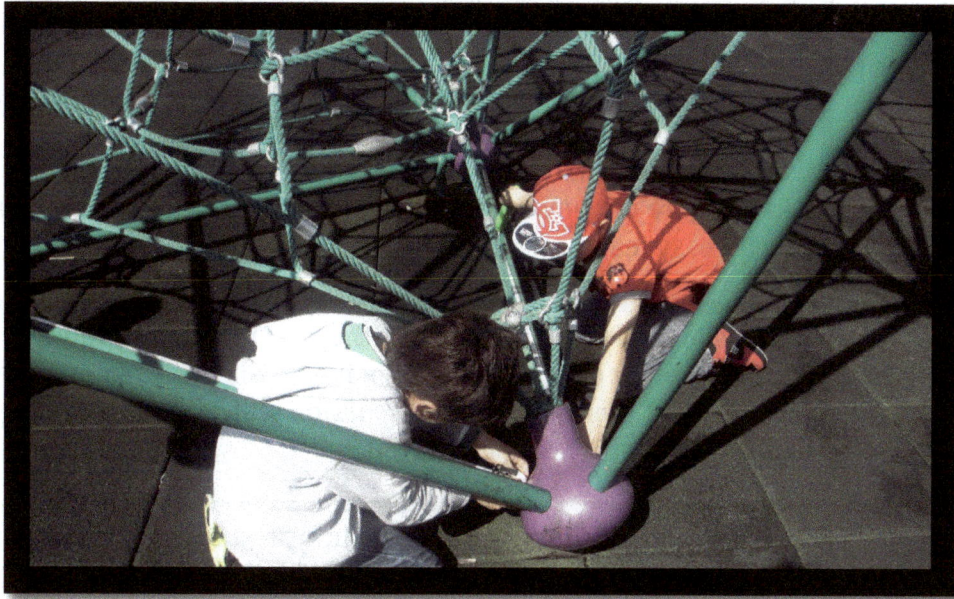

Grade Level: 6

Materials: Flexible measuring tape, geometrically-shaped playground equipment

Objective: In the context of measuring playground equipment, students will experience making a scale model net of the original object.

Note: In our case, we used a piece of equipment with 20 equilateral triangles, thus this lesson has this shape in mind.
The lesson will have to be adjusted based on what kind of equipment that is chosen.

Common Core Connection:

Grade 6
CCSS.Math.Content.6.G.A.4 Represent three-dimensional figures using nets made up of rectangles and triangles, and use the nets to find the surface area of these figures. Apply these techniques in the context of solving real-world and mathematical problems.

CCSS.Math.Content.6.RP.A.3d Use ratio reasoning to convert measurement units; manipulate and transform units appropriately when multiplying or dividing quantities.

Introduction 1 Measuring the equipment	Inform the students that they are going to measure an identified piece of equipment and that they will create a replica of it with construction paper. Students should already be familiar with nets and creating simple three-dimensional figures.
Exploration 1	In small groups, students will go out and measure the equipment. Our piece was a "spider-web" that is constructed with 20 equilateral triangles. The students measured a few key triangles to verify if they were indeed equilateral and that they were all congruent. Then they had to figure out how many triangles made up the construction, which was no easy feat.
Discussion 1	Discussion will be taking place as they are measuring and trying to figure out what type of triangles make up the piece of equipment and how many of them there are. Discussion will be unique to the piece of equipment that is being measured
Introduction 2 The construction of the net	It is now time to construct the net. Inform students that they need to decide on an appropriate ratio. If students are unfamiliar with ratios at this time, take the measurement of one of the sides of the triangles measured and ask them about how big they want their model to be , Then divide the length by different amounts and judge whether or not the quotient is an appropriate length for their model. If the number is divided by ten, for instance, then the model to object ratio will be 1:10.
Exploration 2	Students are to construct the shapes that are needed to make the 3-D model, in this case 20 equilateral triangles. The creation of triangles is very challenging and students should be encouraged to experiment based on what they know about these triangles, 3 equal sides and 3 equal angles of 60 degrees each. My students eventually figured out that the easiest way to draw them was to measure one angle and mark the desired length on those two sides and connect the final side. Still coming up with exact equilateral triangles is extremely complicated and precision is necessary. Constant coaching and mentoring students on an individual level will be necessary.

Discussion 2	The class can discuss the difficulties they experienced when creating the triangles (or other shapes) and which strategies yielded the most precise results.
	Students will need to decide how to finalize their creation by putting together the shapes made. In our case, this was very challenging and required some adult support.
	The final project can be used as a springboard for other learning such as calculating surface area.

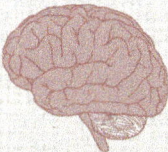

Did you know?   #23

There is evidence that early motor stimulation impacts reading, writing and attention skills. Reading, for example, involves the development and control of smooth eye-movements to send information to the brain, and more importantly, in the right order. Many struggling students are unable to use their eyes to learn and are poor at observing visually as well as having poor organizational skills.

Make it Memorable

Utilize the last 5 minutes of the day to either teach a new concept or remind the brain of essential content taught earlier in the day.

Circumference of Trees & poles

Grade Level: 7th

Materials: 100 ft measuring tape, recording sheet, pencil, cylindrical object

Common Core Connection:

CCSS.MATH.CONTENT.7.G.B.4
Know the formulas for the area and circumference of a circle and use them to solve problems; give an informal derivation of the relationship between the circumference and area of a circle.

Objective: In the context of measuring objects in the environment, students will develop a better sense of estimation of the circumference of cylindrical objects.

Note: You can use this lesson before comparing circumference and diameter and determining the relationship of pi.

Introduction	Show a tin can or fat candle to the group and ask them to estimate what the circumference is. Collect guesses on post-it notes and have them put them on the wall and sort their responses from least to greatest and create a line plot. Measure the object and compare the result to the estimates.
Exploration	Take the students outside and in small groups send them to find poles, trees, or anything else cylindrical to measure. On their recording sheet, they are to first record their estimate (they can write a different one for each student if they don't agree) and then the correct measure. *(table)*
Discussion	Have students discuss what they noticed when they were measuring, such as if they tended to over or under-estimate. Discuss if their estimates were more accurate over time.

Table within Exploration:

Estimate	Actual Measurement

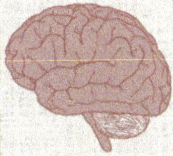

When children have limited experiences positioning their bodies in space and around objects, such as when they are confined to strollers, then they can also struggle with letter orientation and identification a page, as well as geometry and visual-spatial skills.

Make it Memorable

Limit the presentation of new information, especially if cognitively demanding, to no more than 10 minutes.

Balance Beam

Students wondering how
many of this gentleman would fit along the
balance beam.

103

Lesson Name	Grade Level	CCSSI Standard
Balance Beam Fractions	3rd	CCSS.Math.Content.3.NF.A.1
		CCSS.Math.Content.3.NF.A.2
		CCSS.Math.Content.3.NF.A.2a
		CCSS.Math.Content.3.NF.A.2b
		CCSS.Math.Content.3.NF.A.3
		CCSS.Math.Content.3.NF.A.3a
		CCSS.Math.Content.3.NF.A.3b
		CCSS.Math.Content.3.NF.A.3c

Note: Grade level indicates where they appear in the Common Core, however, it may be appropriate to do a lesson in other grade levels depending on conceptual understanding.

Balance Beam Fractions

Grade Level: 3rd

Materials: Balance beam or 12 foot board, large measuring tape (100 foot or one that is at least as long as the board used, access to materials that might be used for measuring (standard and non-standard), index cards to write fractions that can be placed on the beam, chart paper, markers, card stock to create fraction game, hands-on materials that students can use to represent fractions, such as pattern blocks, Cuisenaire rods, double sided counters.

Objective: In the context of building a 0-1 number line on a balance beam, students will develop understanding of fractions as numbers as described in the standards above.

Note: Prior to this activity, students should be comfortable with fractions as being part of a whole and part of a set and know related vocabulary such as numerator and denominator and their relationship to the whole. Thinking about fractions on a number line is quite abstract and can be very difficult to grasp, especially at this age. Constructing can help children make sense of the number line model.

This series will take multiple days. Length will depend on readiness of students and their prior background with other fraction representations.

What is 1/2?

Introduction 1	Review ½: Using "Find a Half" worksheet, have students shade in ½ of each shape as pre-assessment. Discuss as a group the different ways students created a half with each shape.
Exploration 1	Pose to the students the blank "Quilt Squares" and ask students to come up with as many ways to color them to equal ½. Provide different tools such as pattern blocks, Cuisenaire rods, double sided counters, geoboards, etc., and ask students to find as many different ways to model ½.
Discussion 1	Discuss the different models that they created and together come up with a clear definition of ½ and chart, such as: "1/2 is when one object or one group of objects is divided into two equal parts or groups."

Note: When I was working with a small group of struggling students, we spent three 30 minute session just around ½, what it means and how to represent it. Therefore, do not underestimate the importance of really digging into this.

Finding ½ on the Beam (non-standard units)

Note: Work around the balance beam should be limited to approximately 8 students, or it will become too crowded.

Introduction 2	Using the balance beam, 12 foot board, or simply a strip of electrical tape on the floor, that is marked with a "0" and "1" on the ends, ask students to come up with a way figure to figure out where exactly ½ will be so it can be marked.
Exploration 2	In pairs or triads, students will use anything available to them in the classroom to figure out where ½ of the beam is. They will very likely start with trying to count feet, using hand spans, and other things in the classroom. They also may try to just eyeball it. As they are exploring, discuss the difference between estimating and knowing where ½ have is exactly. When students are using tools, they may be creating gaps between iterations.

Discussion 2	Depending on the strategies students tried, discuss the ones that would help them find exactly one half and those that would be considered approximations. For example, if they used a tool that when they laid them out on the beam didn't cover it exactly and with an even number of pieces, then they would not be able to find the exact middle without some kinds of estimations. If they are using their hands to mark where an object ends, then most likely they are creating quite a gap. Mark on the beam the most reasonable middle they come up with so they can compare later when using standard units.

Finding ½ on the Beam (standard units)

Introduction 3	If they have only considered non-standard units, ask them what they could use to know exactly, if there is a tool that could help them out. The goal is to get students to realize that they could use a ruler.
Exploration 3	Allow the partnerships the opportunity to lie out rulers to figure out where the exact middle will be. If the beam or board you are using is 12 feet, then 12 rulers will fit exactly and the middle will be right after the 6th. Again, they need to ensure that they are not creating gaps between the rulers.
Discussion 3	When the exact middle has been determined using standards, then compare with what they came up with non-standard units (if they went that route). One of the main points to emphasize is that ½ is the entire space from the zero up to the point where they marked ½. On a number line, the ½ (or any number) is the point where that much space ends, rather than just a point. Discuss that the point from 0 to the middle is ½ and so is the amount from the middle to 1, however a number line is consecutive and since the portion from the middle to 1 is the second half, its name is "two-halves" or "2/2". Mark 2/2 where the 1 is on the beam. It is important to note that many fraction pieces have fractions, such as ½, written on them in the middle of the piece. Careful distinction needs to be made between the way fraction pieces and number lines are labeled to prevent misconceptions.

Finding fourths on the Beam

Introduction 4	Based on what students understand around what ½ means and how they found ½ on the beam, ask them what they could do to find where ¼ should go. Since students do not use ¼ in their everyday lives in the same way as ½, this will prove to be more difficult. Ask students what ¼ means and perhaps what its relationship is to ½. This will be a difficult question and it might be necessary to pull out some fraction pieces to compare a half and a fourth to ask what they notice. The goal is to get the students to see that ¼ is half of a half as well as know that ¼ means that the beam needs to be divided into four equal pieces.
Exploration 4	Allow partnerships the opportunity to do whatever they need to do to figure out ¼, but encourage standard (rulers), rather than non-standard units. Some groups will try to find the middle of the end to the ½ and others will try to divide the board into four equal parts by laying out the rulers out and then grouping them into 4 groups.
Discussion 4	Have the different partnerships share out their strategy and all the students need to determine which strategy is the most reliable. When a reasonable place has been determined, that spot should be marked with an index card with a ¼ written on it. Ask them what they should call the point after two groups of rulers. Label 2/4 where the ½ is. Ask how much is after the 3rd group and label that ¾, and the same for the entire 4 groups.

Continue this process for thirds, sixths, and eighths.

Most likely, one session will have to be dedicated for each thirds, sixths, and eighths, so that the relationships between fractions can be clearly made.

Culminating the mini-unit

After the balance beam has been completely marked, have students make "Thinking Maps" of each ½, 1/3, and ¼. On a chart paper, write one of the fractions in the center. All students can sit around the circle writing all the fractions that are equivalent to the one written in the center. At this point, students will start to

generalize how to write any fraction that is equivalent to ½. They will say things like, "the numerator added to itself will equal the denominator."

Practicing the number line

Playing the "Fraction Flip" game from the book *Beyond Pizzas and Pies* is a perfect way for students to practice what they have created.   On cardstock, either copy the blackline master from the book, which shows a 0-1 blank number line on one side and the marked fractions on the other, or have students create the number line themselves. To create the card themselves, make sure that the 0 and 1 are on the same spot of both sides of the card. Hint, you can create two number lines, one on top of the other, and fold the paper in half so that one is on the front and one is on the back. The line on one side is blank and the one on the back is marked with the desired fractions and/or decimals.

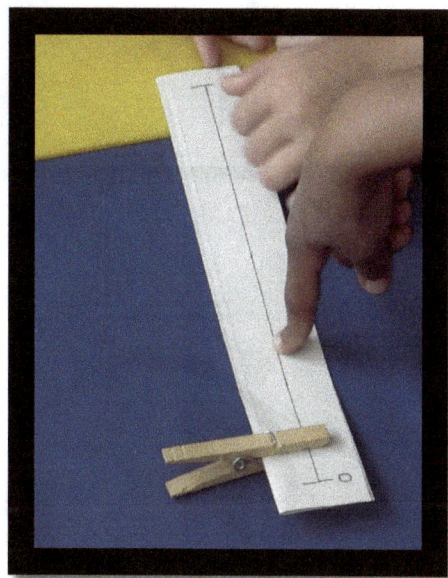

One student draws cards with fractions or decimals written on them and all students must guess where that fraction would be on their open number line. They mark the spot with a clothespin. When all hands are up, have them flip it over to check to see who is closest.

0   1

Find 1/2

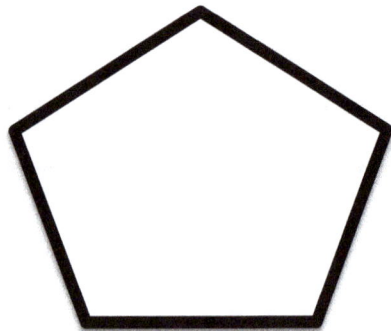

Quilt Squares

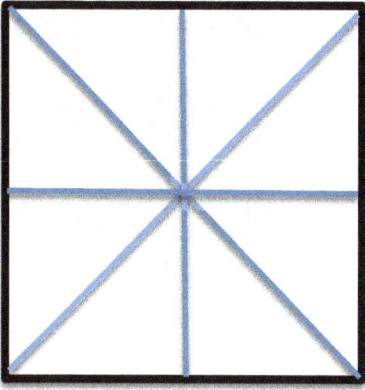

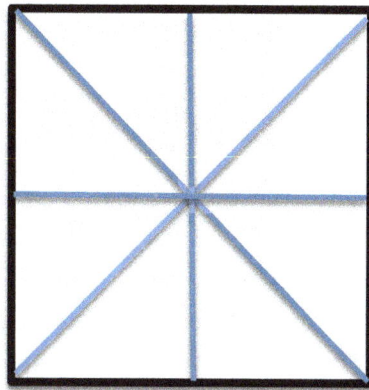

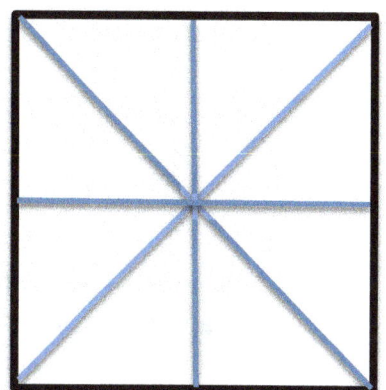

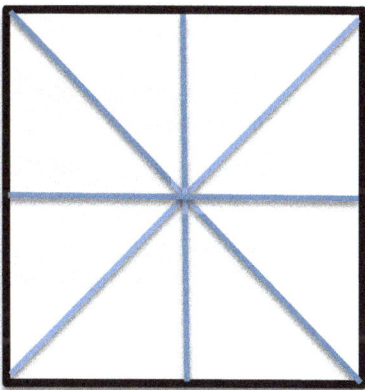

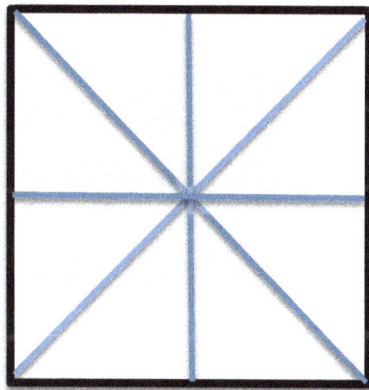

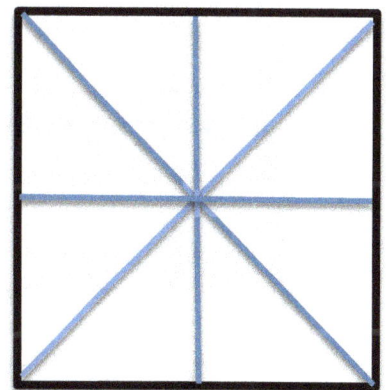

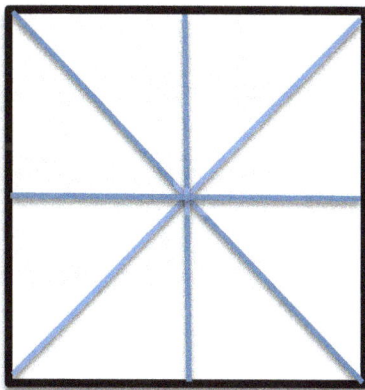

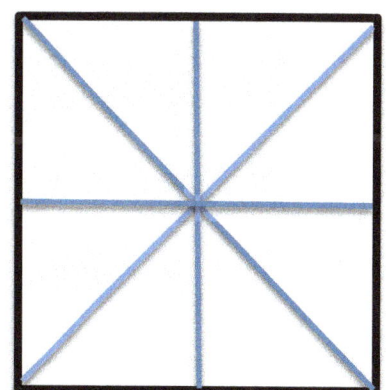

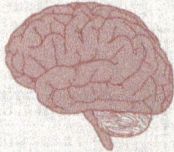

A Teacher's Insight

Even though this lesson falls within the 3rd grade Common Core standards, I find it is a good one for 4th and 5th graders as well, especially if they have not developed a proper foundation with fractions. Operating with fractions will be meaningless if they have not yet built a solid sense of what they are and their relationships to one another.

Rhythm & Music

Lesson Name	Grade Level	CCSSI Standard
The Rhythm Box	Any	Rote Counting and skip counting in a variety of grades
Tap, Tap, Clap, Clap	Any	CCSS.Math.Content.2.NBT.A.2
Measuring Music	3rd	CCSS.Math.Content.3.NF.A.1 CCSS.Math.Content.3.NF.A.2a CCSS.Math.Content.3.NF.A.3b

Note: Grade level indicates where they appear in the Common Core, however, it may be appropriate to do a lesson in other grade levels depending on conceptual understanding.

The Rhythm Box

Grade Level: Any

Materials: A wooden or sturdy step on which students can step up and down, sentence strips, each with a skip counting sequence written on it from bottom to top – include between 12 to 15 numbers in each sequence.

Objective: In the context of using the rhythm box, students will begin to internalize their counting or skip counting sequences to help them learn them with automaticity

Note: Reading numbers from bottom to top is an ideal way to incorporate visual exercises. It helps support eye movements looking up, something that children are not used to doing. Therefore, if the strips are placed higher, they will help work the eye muscles even more.

Common Core Connection:

Rote Counting and skip counting in a variety of grades

Description	Place number strips somewhere in the room, easily visible from the placement of the rhythm box. Using the strip of the number they are working on, students step up and down as they say the numbers in the sequence. They say one number with each step up and nothing on the way down. Note: they are reading from bottom to top.
	Example:
	"3" "6" - - Step up R Step up L Step down R Step down L
	"9" "12" - - Step up R Step up L Step down R Step down L
	"15" "18" - - Step up R Step up L Step down R Step down L
	Alternate rhythm patterns can be created, such as saying the number when stepping up on left and down on right foot.

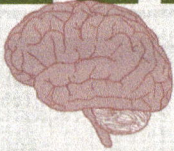

Did you know?

By the 2nd or 3rd grade, the focus of school primarily becomes that of learning information. To be able to do this effectively, the child must have effectively developed his nervous system and have a brain where both halves are working together efficiently. When the brain is unintegrated, the child is unable to cross his eyes or limbs over the midline, an invisible line down the middle of the body. This negatively affects reading and writing. Likewise, the vestibular system, which is connected to the visual and auditory systems and coordinates visual and auditory perception and processing, must be appropriately developed.

Make it Memorable

Try to tie rote memorizing with rhythm or movement. For instance, have students tap their head, tap their shoulders, and snap when learning to count by 3's.

Tap, Tap, Clap, Clap

Grade Level: Any

Materials: Sentence strips used in previous *Rhythm Box* lesson

Objective: In the context of using tapping, clapping and snapping sequences, students will begin to internalize their counting or skip counting sequences to help them learn them with automaticity.

Common Core Standard(s) Addressed:

Skip counting is only briefly mentioned in the common core, however is a fundamental hurdle in learning multiplication tables, especially with meaning. Often, students learn the facts, but do not realize that 3 x 5 and 3 x 6 is only 3 away from each other. Children who have trouble committing facts to memory especially need to go back to learn skip counting, so that they have strategies for figuring out unknown multiplication problems.

Second grade
CCSS.Math.Content.2.NBT.A.2 Count within 1000; skip-count by 5s, 10s, and 100s.

Description	Let students know that they are going to learn how to memorize the skip counting sequence. Choose the number that students need to learn, such as 3. Invite students to choose 3 movements, such as tapping head, tapping shoulders and snapping fingers. As they count silently, they tap head, tap shoulders, then say "3" out loud as they snap their fingers. They may also want to choose tapping a different body part for each number in the entire string, such as head, right shoulder, left shoulder, right hip, left hip, right knee, left knee. Right foot, and left foot. (I found that assigning a number to a body part helps). Teach them that to memorize, it is most helpful to chunk material. In this case, if students are learning their 3's, since most know the first two or three anyway, have them chunk 3 through 18 ("3, 6, 9, 12, 15, 18") and repeat over and over as they engage in the movements described above. When they have this sequence down, start with the last number in the previous sequence (18 in this case) and add on 3 or 4 more numbers (such as, "18, 21, 24, 27"). When this sequence has become more automatic, join it with the previous sequence(s), "3, 6, 9, 12, 15, 18, 21, 24, 27. Continue until all the desired numbers are learned. Extending the sequence beyond 12 numbers may desired. Including the last number from the learned sequence in the new one being learned, acts as a bridge so when they are joined together they will be able to fluidly move from one to the next sequence.

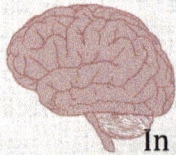

A Teacher's Insight

Students that have a hard time learning their skip counting or basic facts may very well have a hard time memorizing or knowing how to study. In this lesson teaching students how to study is incorporated.

Measuring Music

Grade Level: 3

Materials: chopsticks, fraction bars or strips

Objective: In the context of creating measures of music, students will explore the different ways to create one whole with fractions of ½, ¼, 1/8, and 1/16.

<table>
<tr><td colspan="2">

Common Core Standard Connection:

Grade 3
CCSS.Math.Content.3.NF.A.1 Understand a fraction $1/b$ as the quantity formed by 1 part when a whole is partitioned into b equal parts; understand a fraction a/b as the quantity formed by a parts of size $1/b$.
CCSS.Math.Content.3.NF.A.2a Represent a fraction $1/b$ on a number line diagram by defining the interval from 0 to 1 as the whole and partitioning it into b equal parts. Recognize that each part has size $1/b$ and that the endpoint of the part based at 0 locates the number $1/b$ on the number line.
CCSS.Math.Content.3.NF.A.3b Recognize and generate simple equivalent fractions, e.g., 1/2 = 2/4, 4/6 = 2/3. Explain why the fractions are equivalent, e.g., by using a visual fraction model.

</td></tr>
<tr><td>

Introduction 1

</td><td>

*If you have a subscription to *The Futures Channel* (www.thefutureschannel.com), then show either the clip "Drumming in Fractions" or "The Rhythm Track" and discuss.

Clap some rhythms for children and ask them what they notice (ex: some notes are shorter than others).

Introduce the different notes, its value and how to clap each. Each note should be accompanied by an accompanied speech pattern:
     Whole Note: "Whole Note Hold it,"
     Half Note: "Step Hold"
     Quarter Note: "Walk"
     Eighth Note: "Run"
     Sixteenth Note: "Running"

For example, the following would read, (* signifies where the clap is placed)

</td></tr>
</table>

Whole Note hold it,... Step Hold... Walk... Run... Running
* * * * * *

Discuss with students the equivalence of the different notes to the whole:

"Whole Note Hold It"
*
1

"Step Hold   Step Hold"
*   *
½   ½

"Walk Walk Walk Walk"
*   *   *   *
¼   ¼   ¼   ¼

"Run Run Run Run Run Run Run Run"
*   *   *   *   *   *   *   *
1/8   1/8 1/8 1/8 1/8 1/8   1/8 1/8

Running Running Running Running Running Running Running Running
* * * * * * * * * * * * * * * *
1/16 1/16 1/16 1/16 1/16   1/16 1/16 1/16 1/16 1/16 1/16 1/16 1/16 1/16 1/16 1/16

Exploration 1	On the board, present a combination of notes (perhaps they are pre-cut with construction paper, or programs such as the widget on SMART board can be used to create combinations) and have students attempt to clap to the rhythms. Once they are able to do it successfully, then have them add the walking movements, doing what they say for each note. Tell students that rhythm is the way music moves. The words of speech patterns also have rhythm. Have students, in small groups or partnerships, practice some of the speech patterns you either post on the board or have prepared on a handout.
Discussion 1	Ask what they notice about the relationship between the different notes. They should notice things like: ✓ The faster the note, the more of them that there are; ✓ That two halves equal one whole, two fourths equal one half, etc.
Introduction 2	Using fraction bars or strips, on an overhead or board, have the class help you write a measure equaling four beats. Then have the class clap the rhythm with you.
Exploration 2	Working in pairs, and using the fraction bars provided, students are to create a measure of music by filling out an entire measure with exactly 4 beats. Have them make as many different combinations as possible. Have the partnership and record the combination of fractions made to create one measure using a number sentence on the measure, under fraction pieces. Example: 1/2 + 1/2 = 1 Have students choose their favorite combination and work with another partnership to work together to compose two measures of music, each having four beats. Students will clap their measures using chopsticks and walk to their measures.
Discussion 2	Students will perform their measures for the class and discuss if the measures have the correct number of beats. As a class, discuss each of the beats in terms of fractions.

Note: The following is the blackline master to use with the fraction strips and the student directions, which can be posted on the board or handed out to the partnerships.

Measuring Music

1 – Using the fraction bars provided, create measures of music by filling out an entire measure with exactly 4 beats.

2 – Clap out the measures you created.

Remember, the whole note gets 4 beats, the 1/2 note gets 2 beats, the quarter note gets one beat and the sixteenth note gets a half of a beat.

3 - Record the combination of fractions you made to create one measure using a number sentence on the measure, under your fraction pieces.

Example: 1/2 + 1/2 = 1

4 – Choose your favorite combination and work with another partnership to work together to compose two measures of music, each having four beats. Walk the pattern of your measure and clap them out using chopsticks.

Measuring Music

Measure 1

Measure 2

Measure 3

Measure 4

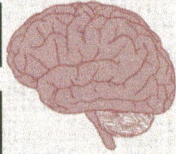

Did you know?

#28

Merzenich discovered that paying close attention is essential to long-term plastic changes in the brain and that neural connections can be altered only if there is full attention to what we do. He noticed that when we try to learn while having a divided attention, the brain maps created are not substantial (Doidge, 2007).

Make it Memorable

Provide comprehensible input by speaking slowly with emphasis and using good body language and gestures, as well as visual supports.

Jump Ropes

Lesson Name	Grade Level	CCSSI Standard
Perpendicular or Parallel	4th	CCSS.Math.Content.4.G.A.1
All Turned Around	4th	CCSS.Math.Content.4.MD.C.5a
Human Pie Chart	3rd-6th	CCSS.Math.Practice.MP4
Jumping contest (Ratios)	6th	CCSS.Math.Content.6.RP.A.3

Note: Grade level indicates where they appear in the Common Core, however, it may be appropriate to do a lesson in other grade levels depending on conceptual understanding.

Perpendicular or Parallel?
(and other geometrical terms)

Grade Level: 4th

Materials: Jump ropes, digital cameras (optional)

Objective: In the context of building parallel and perpendicular lines with jump ropes, students will be reviewing geometrical vocabulary.

Common Core Connection:

Grade 4
CCSS.Math.Content.4.G.A.1 Draw points, lines, line segments, rays, angles (right, acute, obtuse), and perpendicular and parallel lines. Identify these in two-dimensional figures.

Introduction	Group students into groups of 4 and hand each group 2 jump ropes each. Go outside, or a space with enough room to spread out. Shout out either parallel or perpendicular. Students then take their ropes to build what was stated. Have groups justify how they know they made perpendicular or parallel lines. Challenge them by measuring the distance between the two ends of the ropes (for parallel lines) to prove that their lines are indeed parallel.
Exploration	Provide students with a stack of index cards with different geometric terms written on them, such as: parallel, perpendicular, line, ray, right angle, obtuse angle, and acute angle. Groups need to build the vocabulary written on the card and explain as a group how they know that they have correctly built each. They can then take a picture of each one that they have made. Other terms such as shapes that the class is working to learn can be include. It is possible that groups may have to work together to combine ropes.
Discussion	If groups took pictures, then those pictures can be what are shared out. The teacher will pull a card, read the geometrical term from the card and a group can volunteer to share how they built it and how they know that it accurately represents the terminology. The pictures can be either displayed on a word wall or compiled into a class book/dictionary.

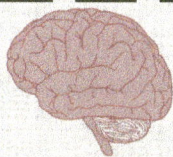

This idea of attention being a necessary agent of change is now being used with many practitioners, such as Anat Baniel. Anat works with children with both physical challenges, such as Cerebral Palsy, and learning challenges, such as Autism. For both of these types of challenges, she states that the number one essential, of the nine that she has coined, that makes the most significant changes is "movement with attention." She states that "the child's ability to notice differences in what she sees, hears, tastes, smells, and feels in her moving body is at the heart of the brain's capacity for creating new neuroconnections and pathways" (p. 30). Therefore, if we want to create changes in our brain maps either on a physical or academic level, significant attention to what we are learning is necessary (Baniel, 2012).

A Teacher's Insight

Is there such a thing as a left angle? When asked, students will generally a right angle as one open to the right and a "left" angle facing left. What does this say about their understanding of a right angle?

All Turned Around

Grade Level: 4th

Materials: Jump ropes, 4-square lines (or perpendicular intersecting lines drawn with chalk), double spinner (one with "clockwise"/"counterclockwise" and the other with the following angle options: 45, 90, 135, 180, 225, 270, 360)

Objective: In the context of playing "All Turned Around," students will review and practice knowing how much to rotate with different degrees.

Common Core Connection:

Grade 4

CCSS.Math.Content.4.MD.C.5a An angle is measured with reference to a circle with its center at the common endpoint of the rays, by considering the fraction of the circular arc between the points where the two rays intersect the circle. An angle that turns through 1/360 of a circle is called a "one-degree angle," and can be used to measure angles.

Introduction	Review with students what an angle is, and in a circle what the major measurements are (90, 180, 360). Ask how to figure out what other amount would be (such as three 90 degree angles, etc.). Introduce the game "All Turned Around", which will be played outside or in a large area, such as gymnasium.

Exploration	During this game, students are broken into partnerships, Each with a jump rope and working over a pair of intersecting perpendicular lines. The teacher will spin the spinners (one which will specify if the students are to go in a clockwise or counterclockwise direction, and the other will specify the number of degrees they have to go). Each student in the partnership will be holding a jump rope at each end. And will be spanning across one of the lines, like this:

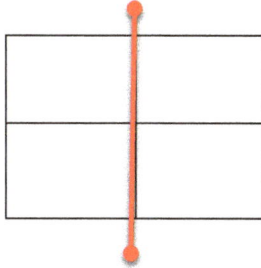

When the teacher says the direction and number of degrees, the partnership, together, has to complete that same turn. For example, if the teacher said "counterclockwise, 90 degrees", then the partnership will rotate to the left and will end up like :

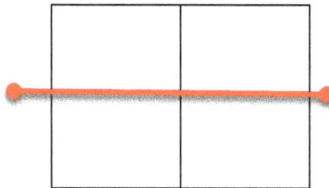

Discussion	When the game is over and the class has returned to their room, they will discuss strategies they had for knowing how far to go with certain degrees, such as 135, 220, 270. The discussion will be based on what the teacher noticed as the problematic areas when conducting the game.

Optional: If it is too noisy for the teacher to be reading aloud the directions or degrees due to being outdoors or in a large room, then each group can have 3 people, one to spin the spinners.

Spinners:

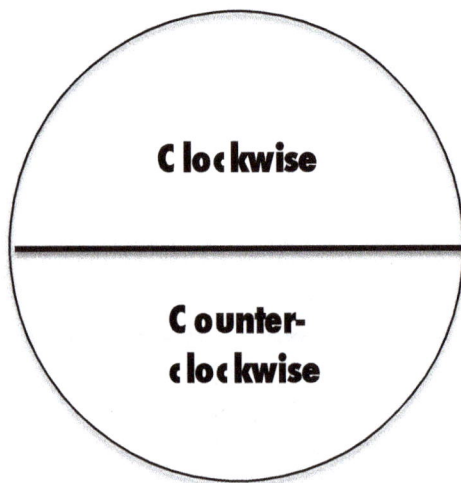

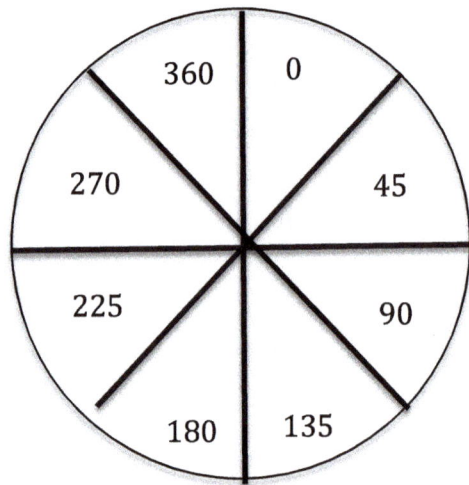

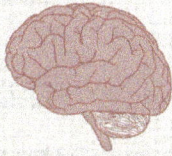

Did you know?

#30

Before appropriate connections have been made in the brain between the senses and structures in the brain, it is difficult for the child to make sense of the word and is very difficult for him to pay attention (Gold, 2008).

Make it Memorable

Encourage students to keep journals or take notes, since the motor skill of writing helps sustain their attention, anchor thought, and memory

Human Pie Chart

Grade Level: 3-6

Materials: Jump ropes or yarn,

Objective: In the context of creating a human pie chart, students will see the relationship between bar graphs and a circle graph, or pie chart.

Note: For this activity, it will be helpful to use data from previous activities from the year, perhaps from beginning of the year activities when getting to know each other.

Common Core Connection:
Although most states included circle graphs (pie charts) in their state math standards, there are no K-8 common core standards that specifically address the use of circle graphs. Creating and interpreting circle graphs would fall in the mathematical practices, for instance, CCSS.Math.Practice.MP4 Model with mathematics

Introduction	Depending on what is being studied in class, choose a question with which to collect data, such as: • How do we get to school: walk, bus, car, bike • What are the demographics of our class: Caucasian, Hispanic, African American, etc. • In this month, the number of days that were: sunny, cloudy, rainy, snowy, etc. • Favorite subject in class: math, reading, writing, science, social studies, art, etc.
Exploration	If the question has to do with the students, have them create a human bar graph (or line plot) – can use the giant 100 grid, then each group of students stick together as the class form a circle. (The teacher can stand in center and will be holding the same number of strings or jump ropes as columns of the graph). The first person from each group will hold the other end of the string or rope. The following graph demonstrates students representing the question of their favorite subject, with choices of science, math, art, writing, and reading. In this illustration there are 26 students.

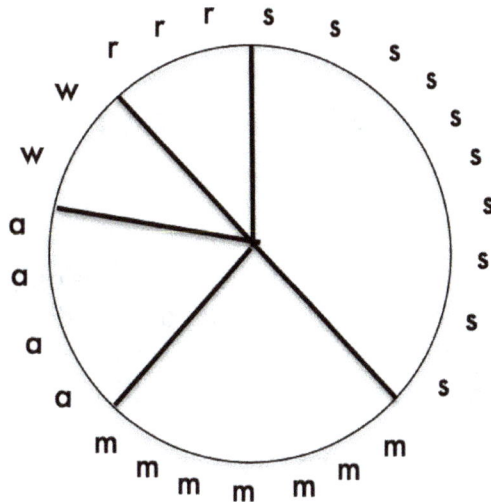

(If possible, take an areal photo). Have students put down the string and step away.

Ask the class to create fractions to represent each section with a fraction (10/26; 7/26; 4/26, 2/26; 3/26). If appropriate for the group, have them reduce fractions that are possible. Students should now convert these fractions to percentages by dividing their fractions.

If appropriate for the grade level, students should create the circle graph on paper of what was created in class, or you can just make a class one on the front board.

Discussion	Invite students to discuss the relationship between the original bar graph and the circle graph. Also, have students discuss what they notice about the circle graph in general, such as that the more students there were in one section, the larger the fraction, or percent. Discuss general difficulties or misconceptions that were noticed during the lesson, especially with drawing the circle graph.

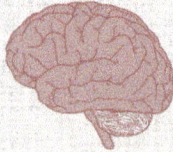

In preschool it is expected that children are only able to pay attention for just a few minutes, however as the child grows, so do the expectations. When children grow in age but their brains are not developing accordingly, they then begin to differentiate from their peers in their ability to attend to the lessons.

A Teacher's Insight

Throughout the year, collect different kinds of data from your students or something meaningful to them. Keep it in a data book so that you can draw from it for lessons such as these.

Jumping Contest

Grade Level: 6th

Materials: Jump ropes, enough for the number of small groups, timer or clock with second hand

Objective: In the context of having a jumprope contest, students will determine whose jumping rates are the fastest and slowest.

Common Core Connection:

Grade 6
CCSS.Math.Content.6.RP.A.3 Use ratio and rate reasoning to solve real-world and mathematical problems, e.g., by reasoning about tables of equivalent ratios, tape diagrams, double number line diagrams, or equations.

Note: To be able to do this lesson, students need to be familiar with how to compare unlike fractions.

Introduction	Choose two students to come to the front of the class and jump rope. Time each for 30 seconds and have two different students count the number of jumps for each jumper. Write the scores as ratios up on the board and ask students to compare which rate was faster and how they knew. Since the time was the same for each, it will be easy to compare. Now see what the unit ratio is for each student. For example, if someone did 6 jumps in 30 seconds, then the unit ratio would be 1 jump in 5 seconds, or 1:5.
Exploration	Inform the class that they will all be jumping and collecting data, but not all will be timed the same way. Each member of the group will be jumping a different length of time. They will then compare and rank the jumps from fastest to slowest. Divide the class into small groups of 3 students each. In each group, label students either A, B, or C. First the A's are going to jump and the C's are going to be the time-keeper. When A jumps, he will jump for 15 seconds, Student B will count the number of jumps and student C will keep time. When student B jumps, he will jump for 20 seconds, student C counting and student A keeping time. When student C jumps, he will jump for 25 seconds. Student A will count and student C will keep time.

	Have students rank their jumps and show their work on how they determined who jumped fastest, etc. Have the groups add data to class charts.
Discussion	As a class, students will look at the fractions in the 15-second column and rank them in order. Some groups may have recorded them as unit fractions, so the class will have to justify their ranking. Do the same for 20 and 25 seconds. Now the class will take their top winners and compare them (if there are 6 groups, they will be comparing 6 students' rates). This will be considerably more difficult because the winner could have been from either of the three categories.

Jumping Contest
Data Sheet

15 sec	20 sec	25 sec
4/15	7/20	8/25

Group Ranks

Group A _____
             _____
             _____

Group B _____
             _____
             _____

Group C _____
             _____
             _____

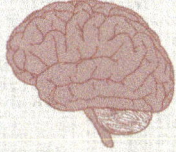

Did you know?

The brain is not designed for continuous learning, but needs some down time to process and "digest" what it is trying to learn. Throughout the day, our brain has natural lows and highs. Each cycle lasts approximately 90-110 minutes, and constant focused attention can only last approximately 10 minutes (*The Brain-Movement Connection,* n.d.). Therefore, in the classroom, teachers need to present essential material in short segments and then allow time for processing, whether individually or in small groups. Fatigue is a sure way to turn off the child's learning switch. Frequent mental breaks are critical for increased learning and productivity.

Make it Memorable

Concepts before procedures. Make sure that students are not trying to commit to rote memory what they do not yet understand. When there is deeper understanding around a concept, the related facts and procedures are easier to memorize. Also, trying to put a procedure to memory without understanding can cause students to learn incorrectly, and thus creating the "bad habits" that are hard to undo.

Odds & Ends

Lessons that Use Commonly Used or Home Made Items

Lesson Name	Grade Level	CCSSI Standard
Place Value Jumping	Any	CCSS.Math.Content.2.NBT.A.3 CCSS.Math.Content.4.NBT.A.2 CCSS.Math.Content.5.NBT.A.1 CCSS.Math.Content.5.NBT.A.3
Hop Scotch Facts	Any	Number and operations in any grade
Venn Diagrams	Any	CCSS.Math.Practice.MP4 CCSS.Math.Practice.MP5
Treasure Hunt	Any	CCSS.MATH.CONTENT.1.OA.D.7 CCSS.MATH.CONTENT.5.OA.A.2
Get the Picture?	Any	CCSS.Math.Practice.MP4 CCSS.Math.Practice.MP7
Human Equations	K-1	CCSS.MATH.CONTENT.1.OA.D.7
Body Shapes	1st	CCSS.Math.Content.1.G.A.2
Architecting Polygons	3rd	CCSS.Math.Content.3.G.A.1
Parachute Rotation	4th	CCSS.Math.Content.4.MD.C.5a
Paddle Up!	6th	CCSS.Math.Content.6.RP.A.3
Tangram Proportions	6th	CCSS.Math.Content.6.RP.A.1
Base in Balloons	5-8	CCSS.Math.Practice.MP7 CCSS.Math.Practice.MP8
Rock, Paper, Scissors	7th	CCSS.Math.Content.7.SP.C.5 CCSS.Math.Content.7.SP.C.6
Hula Hoop Circumference	7th	CCSS.Math.Content.7.G.B.4

Note: Grade level indicates where they appear in the Common Core, however, it may be appropriate to do a lesson in other grade levels depending on conceptual understanding.

Place Value Jumping

Grade Level: 2nd – 5th grades

Materials: Sentence Strips (color if possible), markers, tape, plenty of base ten materials, especially flat (100) and 1,000 cubes.

Objective: In the context of creating the place value charts, students will discover patterns inherent to the chart. Also, by jumping up and down the place values, students will practice knowing the order of the places in ascending and descending order.

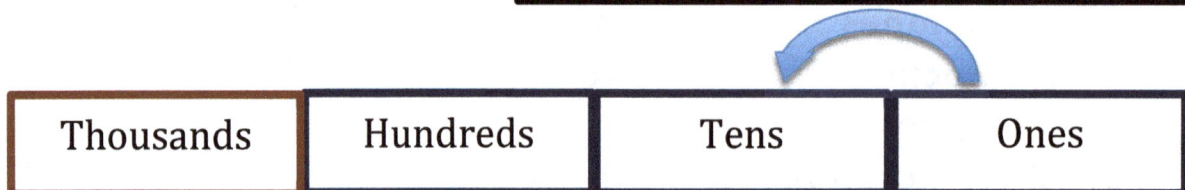

| Thousands | Hundreds | Tens | Ones |

Introduction Building the chart	Show students one cube from the base ten blocks. Ask how many they see. Remind them that when they have 0-9 of the cubes, the digit is in the ones place. Write the word "Ones" on a sentence strip in large letters that can be seen from a distance. (depending on the age range of the students, the number "1" can be written underneath. Also write the word "cube" somewhere on the strip. Inform students that as a class they are going to be building the place value chart up to the value that they are studying in class.

Exploration	Ask students what would happen once they had more than 9 cubes. Students should know that they make a group, either in a cup, making a train or trading for a rod. Ask them what that new group is called. Have someone make a sentence strip that says "Tens," write the number 10 underneath, 1 x 10 and the word "rod." Ask students what it would look like if there were ten of the rods. They should recognize that it would make a 100 flat (build to prove). Have someone make a sentence strip that says "hundreds" with 100 written underneath, 10 x 10, and "flat". Ask students what 10 of those flats would look like. Some may have seen a 1,000 cube, or just stack them and notice it makes a cube. Others may just line them up into a train, but highlight the cube that is created. Ask how many are represented now. (Just providing a commercial 1,000 cube is deceiving because it is hollow and students count the 6 faces instead. By stacking the flats, they count and realize that it is worth 1,000). Have someone create a sentence strip with "thousands" written on it, the 1,000 number as well as 10 x 10 x 10, and the word "cube" (since they made a larger cube). Students are generally unfamiliar with any piece that is created with 10 of these 1,000 cubes. Ask them what it would look like if they had 10 of them and to build it. You may see different configurations, but highlight the model that shows a train of ten of the large cubes, otherwise a rod. Ask how many are represented, then have someone write "ten-thousands" on a strip, 10 x 10 x 10 x 10, and "rod". Continue this procedure until the desired place is reached, such as billions (*note: if building a million, only a skeleton will be able to be created due to a lack of base ten blocks, but students will get the idea that it makes a large cube, and the dimension is one cubic meter, so meter sticks can be used to construct it*). *Decimal numbers may be included at this time, or perhaps added on when decimals are studied in class later in the year. When doing so, start with a base ten piece that will be able to be broken down into the number of decimal places desired. Always start with a cube, since the cube (one) precedes the tenths place. For example, start with the thousand cube made up of 10 flats. Ask students what would happen if you cut the whole into ten equal parts, how much would you have (1/10 of the whole, or a flat). On the sentence strip write "tenths", 1/10; 1×10^{-1}, and flat. That flat can be further divided into ten equal pieces (1/100 - rods) and then again (1/1000 – cube).*

Discussion	Since this activity is done as a whole class, undoubtedly there will discussion throughout the entire activity. However, once all the sentence strips are created, line them up wherever they are going to be placed (preferably along the floorboard, maybe in a hallway). Ask students what they notice about the entire chart.
	
	Some things you want to surface are:
	
	   ✓ There is a pattern of the shape that was made (cube, rod, flat).
	   ✓ Emphasize that these three make up a "period" in the place value system and that they repeat over and over, even in the decimal numbers (sentence strips of the same period should be the same color for emphasis)
	   ✓ Connect to exponents: $10 \times 1 = 10^1$; $10 \times 10 = 10^2$; $1/10 = 10^{-1}$
	   ✓ The exponent shows how many zeros there are in the number

Practice Jumping	Periodically allow students to jump on the place value chart, either starting with one, the largest place, or smallest place and saying the name of that place. By jumping in different directions and saying the order in ascending and descending order, students start to internalize the order. When children are learning to read large numbers, if they cannot remember the places, I send them to the place value wall to jump the places. You can also give cues such as "x 10," "x 100," "÷10," and have them jump the corresponding places.

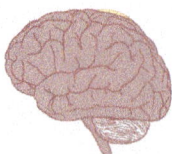

One way to appropriately attend is to go slow, whether it is a movement or thinking about a new math concept. Fast, we can only do what we already know, and movement done automatically creates little or no new connections in the brain. In fact, when we do things quickly, the brain defaults to "already existing and deeply grooved patterns" (Baniel, 2012).

A Teacher's Insight

This lesson will vary depending on the grade level. This lesson will need to take place after some introductory work to the place value system and serves as a practice component. It served me well with my older students who just could not remember the different places. They were really able to connect multiplying/dividing 10, 100 and 1,000 to moving the decimal point and were able to mentally calculate much easier.

Hop Scotch Facts

Grade Level: Any

Materials: Electrical tape, index cards with numbers (taped in the squares).

Objective: In the context of playing *Hop Scotch Facts* students will practice mentally calculating.

| Description | Create a grid on the floor using electrical tape. The grid doesn't have to have a specific shape, but a 5 x 5 grid would be an example. Inside each square have a number taped to the floor. The size and kind of the number (i.e., fraction, integer, single digit, etc.) would depend on what is being studied in class.

Pick a number with dice, a spinner, pulling a card, etc. Students will have to jump on numbers that will equal the desired number. Students are welcome to use any operation and need to say the number sentence as they jump, but they need to jump on a minimum of two numbers to make it.

For example, the number chosen was 13. Students might jump on a 6, 3, 5, saying, "6 x 3 – 5." |
|---|---|

21	22	23	24	25
16	17	18	19	20
11	12	13	14	15
6	7	8	9	10
1	2	3	4	5

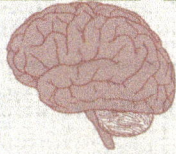

Did you know?   #34

When we do things fast, or with automaticity, it is the brainstem at work. However, when we do things with attention, connections in the brain are being made. Have you ever driven home and not remembered how you got there? That was your brainstem at work. When trying to learn a new skill, it is important to hold off on going fast until the brain has formed the necessary connections and patterns for performing the new skill (Baniel, 2012).

Make it Memorable

Allow students to stand up and walk around during a lecture . Too much sitting results in "poor breathing, strained spinal column and lower back nerves, poor eyesight, and overall body fatigue,," (Jensen, 2000, p. 34).

Venn Diagrams

Grade Level: Any

Materials: Large Venn Diagram created with tape, yarn, or sidewalk chalk

Objective: By actively being a part of the Venn diagram, students will understand sets, unions, and intersections, as well as use them to solve mathematical problems.

Common Core Connection

CCSS.Math.Practice.MP4 Model with mathematics.
CCSS.Math.Practice.MP5 Use appropriate tools strategically.

Note: Venn diagrams can be used in virtually any grade, however, the examples provided below are more appropriate for upper elementary, about 4th -6th grades. They can easily be adjusted to meet the needs in the lower levels as well.

Introduction	More than likely, most of your students have some experience with Venn diagrams, perhaps in literacy or social studies. Inform students that today they will use them to solve some math problems as they explore the parts of the diagram and their terminology.
	On the board, write: *How we get to school*. Ask students how they get to school in the morning and write all the different ways on the board. Choose two of the most common ways that you know students in your class arrive to school, such as riding the bus and walking. Lead the students to the area that you have prepared with a giant Venn diagram taped to the floor. On the outside of each circle, label it with what they should represent (perhaps with sidewalk chalk, or tape a paper to the floor). One by one, ask students whether or not they ride the bus, walk, or used a different mode of transportation. If they ride the bus and never walk, then they need to stand in the "bus" circle. Likewise, if they walk and never ride the bus, they can stand in the "walk" circle. If they use another form of transportation, such as a bike or ride from parents, then they stand outside the diagram, and if they sometimes walk and sometimes ride the bus, then they stand in the center.

	Ask students to identify what is the **set**, or the commonality between all the groups. Inform them that the **intersection** is the group that are a part of both circles, or in this case those that walk and ride the bus, and the **union** is the students that are in either group, but not both, and the **difference** is the students that are in just one specific group. Continue the activity with the entire class for different sets, such as: • Number of students with brown hair and brown eyes • Number of students who brought home lunch and drank milk for lunch
Exploration	Step 1: Write the numbers 1 to how ever many students are in the class, written separately on a index card and pass them out to each student. On the Venn diagram taped to the ground, label each circle as shown below: Multiples of 2 Multiples of 3 Have each student, one by one, go to their appropriate spot. Ask students to identify which numbers are in the union, intersection, and difference. Step 2: In small groups, students will create a Venn diagram on paper, large enough to be seen by others if hung on a wall. Students will be given two numbers, such as 24 and 30 and their task is to use the Venn diagram to come up with the GCF (Greatest Common Factor). To do this, students will have to identify the prime factorization of each number and the ones that are shared by both will go in the intersection. Those factors that are unique to each number will be placed in its appropriate circle. The factors in the intersection are multiplied to each other to determine the GCF.

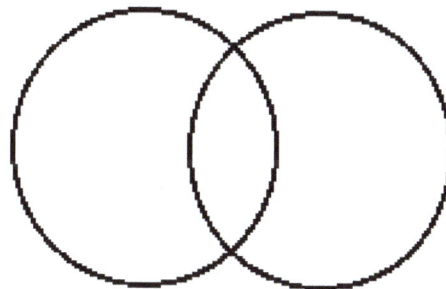

Discussion	Groups are selected to share out their Venn diagram and which numbers are in which category, using appropriate vocabulary. They are also to discuss this way of identifying GCF in comparison to other ways that they have learned in class.

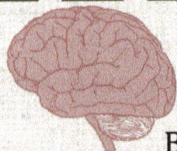

Did you know?   #35

Baniel (2002) explains that when one goes slow, it allows the brain a chance to feel. Einstein came up with his theory of relativity by imagining himself riding a ray of light, feeling the sensations of movement and the relationships of his body to the space around him. Many children do not know how to go slow and attend and notice the nuances in what they are doing. It is up to us at teachers to encourage and coach students to slow down and to notice relationships and patterns to help develop a deeper conceptual understanding.

A Teacher's Insight

Graphic organizers can really help students organize and see relationships between ideas, whether in math or literacy.
Thinking Maps® is a collection of a variety of graphic organizers that I have witnessed make powerful impacts on children's thinking, especially in the younger grades. http://thinkingmaps.com/

Treasure Hunt

Grade Level: Any

Materials: Clue cards with equations written on them (such as 50÷5) written on them

Objective: In the context of playing the game *Treasure Hunt,* students will have the opportunity to practice their facts or mental math strategies.

Common Core Connection:

Grade 1
CCSS.MATH.CONTENT.1.OA.D.7
Understand the meaning of the equal sign, and determine if equations involving addition and subtraction are true or false.

Grade 5
CCSS.MATH.CONTENT.5.OA.A.2
Write simple expressions that record calculations with numbers, and interpret numerical expressions without evaluating them.

Description	All students are provided with a clue card with a number sentence written on it. It is their job to figure out who has an equation that is equivalent to theirs without shouting out the answer to their own problem. First, they will need to solve their own equation, then as they roam the room, they need to solve equations of their classmates and find one that is equivalent to their own.

9 + 2

10 + 1

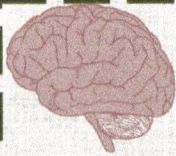

"Learning is easier to store, remember, and retrieve if it has an emotional base," (Oberparleiter, 2004, in Lengel, & Kuczala, 2010, p. 19). Students who feel safe and respected are more highly motivated to learn and are more likely to be intrinsic learning. The limbic system, which is involved in our emotions and hormone control, is comprised of several parts of the brain, including the amygdala and hippocampus, which play an important role in memory.

Make it Memorable

Some students need to have something in their hands to focus and concentrate. Allow students to use fidgets and doodle if necessary.

Get the Picture?
Using the Environment for Mathematical Representations

Grade Level: Any

Materials: Digital Camera

Common Core Connection:

CCSS.Math.Practice.MP4 Model with mathematics.
CCSS.Math.Practice.MP7 Look for and make use of structure.

Objective: In the context of getting out and taking pictures, students will demonstrate understanding of the mathematical concepts and vocabulary being taught in the classroom.

Description	Getting out in the environment can help math come alive for students. When learning about math concepts and vocabulary, whether in geometry, fractions, measurement, etc., invite students to grab a digital camera and get out in the open to find real-life examples of what is presented in the classroom. Have students create books or math word walls with their pictures, written descriptions, and numerical representations to match. Examples: ✓ Parallel and perpendicular lines ✓ Polygons ✓ Right, obtuse, acute angles ✓ Objects that are longer or shorter than one foot ✓ Something that is arranged as an array (include multiplication sentence to match) ✓ Something to demonstrate ½ (or any fraction).

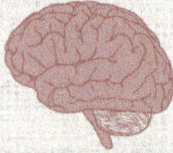

The limbic system is stimulated by the Reticulating Activating System (RAS), which is then connected to the pre-frontal cortex by dopamine (Blomberg & Dempsey, 2011). The thalamus is also a structure of the limbic system, which is involved in sensory perception and regulation of movement. Negative emotions not only switches off learning, but can cause the release of neurotransmitters that can weaken the immune system, therefore having physiological affects. Therefore, there is a strong connection between our emotions, memory, and movement.

A Teacher's Insight

When students are invited to explore and notice geometry in their environment, it is likely they will act like they are seeing the world for the first time. It is precious to watch them notice things they nave never seen before. They get so excited and start to see geometry everywhere.

Human Equations

Grade Level: k-1

Materials: Large number cards to hold, large equal sign taped to the back of a chair, pan balance, a small equal sign to be taped in the center of the pan balance.

Objective: In the context of acting out equations, students will deepen their understanding and be able to articulate what the equal sign represents and practice the idea of more and less.

<table>
<tr><td>Common Core Connection:

Grade 1
CCSS.MATH.CONTENT.1.OA.D.7
Understand the meaning of the equal sign, and determine if equations involving addition and subtraction are true or false.</td></tr>
</table>

Note: This lesson can be modified to fit other grade levels. In Kinder and first grades, this most likely will be a repeated lesson over multiple days, just changing the numbers. This is a whole group lesson.

Introduction	Using a pan balance, ask what happens when one cube is placed on one side and two cubes are placed on the other. Ask what they can do so that the balance can stay straight. Students will suggest either adding a cube to the side with only one, or perhaps taking away a cube from the side that has 2. Choose one of their suggestions, and try it out. Allow them to notice that the balance is straight. When the balance straightens out, ask students what they notice is happening. Suggest that one side is not more than the other, so the balance is now balanced, or equal. Tell students that they are going to be like the cubes on the balance. Put a chair in the center with an equal sign taped to the center. If you want, you can even tape a large circle on either side of the chair to act as the two sides of the balance.
Exploration	Select 3 students from the class to come and stand on one side of the chair (or balance). Give the group a "3" number card to hold. Then select 2 students to come and stand on the other side of the chair and provide them with a "2" number card to hold. Ask the class which side is more and which side is less. Have the students that are more sit down to act like their side of the balance is down. Now ask the remaining students to talk with a partner to decide what they should do to make the two sides equal.

Discussion	After sufficient time for discussion has passed, ask students for suggestions and act them out to see if they work. On the board record their ideas using numerical representations. For example, if students suggest taking away 1 from the side with the 3, then write:

$$3-1 = 2$$

This will help kinder students connect to more abstract representations. First grade students can be encouraged to both record pictorially what is being acted out and to represent that with a number sentence.

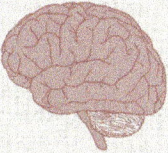

Did you know?

The patterns formed in the child's "brain will take in everything he experiences while doing exercises - physical or cognitive - including patterns of not being able to perform that movement or skill or not being able to do it well" (Baniel, 2012 p. 52). What this means to teachers, is that as students are focusing on trying to learn a skill, if there is negative emotion, stress, and frustration attached to that learning experience, it will be wired together with that skill he is trying to learn.

Make it Memorable

Provide a variety of tools and manipulatives for students to use and encourage them to make connections between them.

Body Shapes

Grade Level: 1ˢᵗ Grade

Materials: Camera (optional)

Objective: In the context of creating shapes with their bodies, students will internalize how shapes are and feel.

Common Core Connection:

Grade 1
CCSS.Math.Content.1.G.A.2 Compose two-dimensional shapes (rectangles, squares, trapezoids, triangles, half-circles, and quarter-circles)

Introduction	Review the names of common shapes. Create small groups of about 3 students and tell them that together they will make shapes with their bodies. They need to use all the people in the group to make the shape.
Exploration	Randomly call out a shape and have the groups create it. They can be standing or lying on the floor, so make sure there is an appropriate surface and space to lie down. Ask groups how they know they have made the desired shape. Take pictures if possible to add to a vocabulary wall.
Discussion	If pictures were taken, they can be shown up on the wall and the entire class has to identify what shape had been made and how they know. If pictures were not taken, have a group volunteer to come up and model the shape and defend how they know it is correct. OR have a group secretly pick a shape to make and have their classmates guess what they made.

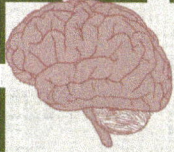

Did You know?   #39

Short-term memory includes both immediate and working memory, each with very restrictive time and capacity limits (Sousa, 2008). The hippocampus uses "sensory input from the thalamus, movement coordination in the basal ganglion and emotions in the hypothalamus to form short-term memory. Communication between hippocampus and the brain area that handles sensory information fortifies our memories" (Hannaford, 1995, p.60).

A Teacher's Insight

Children who experience problems with perception and directionality would benefit from activities such as crawling through and exploring the inside and outside of a variety of 3-D figures, such as a rectangular prism (refrigerator box), tube, pyramid (t-pee or tent), and triangular prism (modified appliance box).

Architecting Polygons

Grade Level: 3rd

Materials: Long strands of yarn

Objective: In the context of creating polygons with themselves within strands of yarn, students will be able to articulate what properties make a polygon.

Common Core Connection:

Third grade
CCSS.Math.Content.3.G.A.1 Understand that shapes in different categories (e.g., rhombuses, rectangles, and others) may share attributes (e.g., having four sides), and that the shared attributes can define a larger category (e.g., quadrilaterals). Recognize rhombuses, rectangles, and squares as examples of quadrilaterals, and draw examples of quadrilaterals that do not belong to any of these subcategories.

Introduction	Provide groups of about 5 students with long strands of yarn, a couple of meters long, tied together at the ends. Shout out "quadrilateral" and expect the students to create this shape while inside the yarn (students will be acting as the vertices). In come cases, students may have to stand out of the polygon since the number of vertices is less than the number of people, or they may get creative with how they incorporate everyone.

	Ask groups to justify how they know they have made the quadrilateral. If they all made a rectangle, then challenge them to come up with as many different ways to make a quadrilateral. Hopefully there will be figures such as: squares, parallelograms, kites, cheverons, etc.
Exploration	Give each group a stack of cards, each with a different polygon written on it. Ask the groups to pick a card and to make it in as many different ways as possible. If possible, pictures can be taken by one member of the group. Continue with several of the cards.
Discussion	When the groups are done, ask for volunteers to show how to make selected polygons. Ask students how they knew that they had correctly formed it. Students should be noting the number of vertices or sides. The polygons that require more vertices than people will be challenging to them.

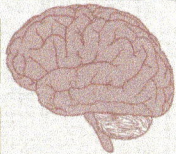

Did you know? #40

During a lesson, we tend to remember best which comes first and second best that which comes last (Sousa, 2008). This is known as the *serial position effect.* The reason this happens is that the information in the first few minutes is within the working memory's capacity limits, but later during the lesson, our capacity limits have been exceeded. The following diagram by Sousa, 2006, illustrates what this might look like in a 40 minute lesson. The retention time periods are labeled as *prime-time-1* and *prime-time-2.*

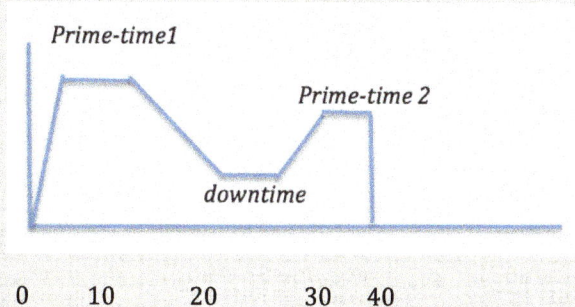

Prime-time1

Prime-time 2

downtime

0   10   20   30   40

The degree of retention during a learning episode (Sousa 2008, p. 61)

Make it Memorable

Use Games as a way to learn material. It not only increases motivation, but the positive emotion that is generated will be what is wired together to the information being learned.

Parachute
Rotation

Grade Level: 4

Materials: Parachute (used in physical education)

Objective: In the context of using the parachute, students will review and practice degrees of rotation.

Note: This lesson is similar to "All turned around" in the previous section.

Common Core Connection:

Grade 4
CCSS.Math.Content.4.MD.C.5a An angle is measured with reference to a circle with its center at the common endpoint of the rays, by considering the fraction of the circular arc between the points where the two rays intersect the circle. An angle that turns through 1/360 of a circle is called a "one-degree angle," and can be used to measure angles.

Introduction	Review with students what an angle is, and in a circle what the major measurements are (90, 180, 360). Ask how to figure out other amounts (such as three 90 degree angles, etc). Introduce the activity and the parachute. It might be wise to allow the class to experiment with the parachute before beginning the activity.
Exploration	Students will be standing on the circumference of the parachute, holding on to the side. Make sure that students are evenly spaced. The teacher will spin the spinners (one which will specify if the students are to go in a clockwise or counterclockwise direction, and the other will specify the number of degrees they have to go). When the teacher says the direction and number of degrees, the class has to complete that same turn. For example, if the teacher said "counterclockwise, 90 degrees", then the class will rotate to the left .
Discussion	When the game is over and the class has returned to their room, students will discuss strategies they had for knowing how far to go with certain degrees, such as 135, 220, 270. The discussion will be based on what the teacher noticed as the problematic areas when conducting the game.

Spinners:

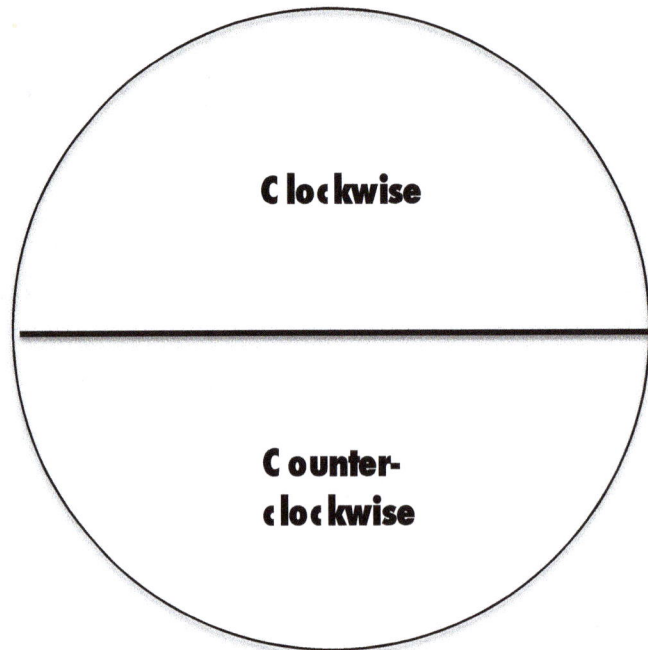

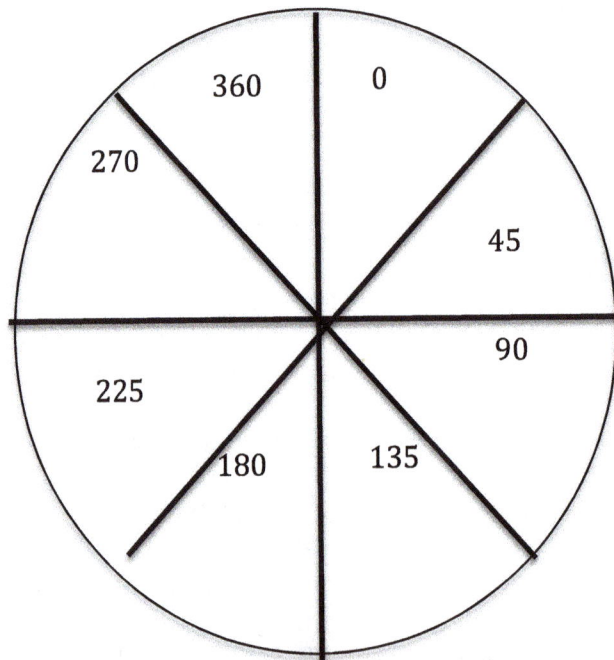

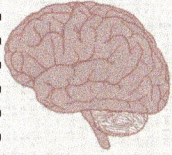

A Teacher's Insight

When coming up with lessons to do with different equipment, I watched students engaged in stations at a PE fair at the local university. I looked for the geometry and number that might be inherent at each station. Coming up with lessons can be as simple as identifying a tool or piece of equipment and thinking of all the things that can be done with it and find where the math might be.

Paddle Up!

Grade Level: 6th

.Materials: paddle or racket, ball, such as a ping pong or tennis ball, and timer or clock with second hand

Objective: In the context of having a contest of hitting a ball with a paddle, students will determine whose paddling rates are fastest and slowest

Common Core Connection:

6th grade
CCSS.Math.Content.6.RP.A.3
Use ratio and rate reasoning to solve real-world and mathematical problems, e.g., by reasoning about tables of equivalent ratios, tape diagrams, double number line diagrams, or equations.

Note: This activity is similar to the *Jumping Contest* activity

Introduction	Choose two students to come to the front and hit a ball with a paddle or racket. Time each for 10 seconds and have two different students count the number of hits for each student. Write the scores as ratios up on the board and ask students to compare who had the most hits in the ten seconds. Since the time was the same for each, it will be easy to compare. *Note: students can pick the ball and keep going if they drop it.* Now see what the unit ratio is for each student. For example, if someone did 20 hits in 10 seconds, then the unit ratio would be 2 hits in 1 second, or 2:1.
Exploration	Inform the class that they will all be paddling and collecting data, but not all will be timed the same way. Each member of the group will be paddling for a different length of time. They will then compare and rank the paddles from fastest to slowest. Divide the class into small groups of 3 students each. In each group, label students either A, B, or C. First the A's are going to paddle and the C's are going to be the time-keeper. When A paddles, he will do so for 15 seconds, Student B will count the number of hits and student C will keep time. When student B paddles, he will do so for 20 seconds, student C counting and student A keeping time. When student C paddles, he will do so for 25 seconds. Student A will count and student C will keep time. Have students rank their hits and show their work on how they determined who had the most hits. Have the groups add data to class charts (see next page)

Discussion	As a class, students will look at the fractions in the 15-second column and rank them in order. Some groups may have recorded them as unit fractions, so the class will have to justify their ranking. Do the same for 20 and 25 seconds.
	Now the class will take their top winners and compare them (if there are 6 groups, they will be comparing 6 students' rates). This will be considerably more difficult because the winner could have been from either of the three categories.

Paddling Contest
Data Sheet

15 sec	20 sec	25 sec
4/15	7/20	8/25

Group Ranks

Group A _____
         _____
         _____

Group B _____
         _____
         _____

Group C _____
         _____
         _____

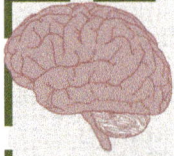

Make it Memorable

Make the mathematics life-size. If a problem can be manipulated on a worksheet, it can become giant-sized where the students become the pons. The majority of the lessons in this book demonstrate this idea.

Tangram Proportions

Grade Level: 6th Grade

Materials: Sets of Tangrams, poster board, rulers, markers

Objective: In the context of re-creating tangrams of a larger scale, students will understand how to use multiplication and division to enlarge or shrink proportionally.

<table>
<tr><td colspan="2">Common Core Connection:
Grade 6
CCSS.Math.Content.6.RP.A.1 Understand the concept of a ratio and use ratio language to describe a ratio relationship between two quantities.</td></tr>
</table>

Introduction 1	On the computer screen, or demonstration board, show a picture. Ask students what would happen if the length of the picture increased, but the width did not. Show by dragging the picture to the right in a straight line. Now ask the students what they need to do so that the picture remains proportional, or so that it does not look stretched.
Exploration 1	Give students a I" square tile. Ask them to measure the tile (to confirm that it is 1" all the way around) and how big they would need to draw it if they wanted it to be 3x as big (or a 1:3 ratio). Have students draw the square.
Discussion 1	Discuss difficulties. Let them know that since the picture and object should be proportional, the angle measurements will be the same and they might want to use the corners of the squares to help them in their drawings
Exploration 2	Now provide students with one of 7 pieces of the tangram puzzle. Have the class choose what they want their ratio to be, not too small, yet not too large. They might have to test out their theories ahead of time to make sure that the size would be adequate. They are going to be drawing their piece on poster board. When they are done, they are to get together with the other 6 people with tangrams pieces from their set and put their puzzle together. They should fit together nicely.
Discussion 2	Share out the fitted tangrams and discuss the pitfalls such as maintaining the correct angle measurements.

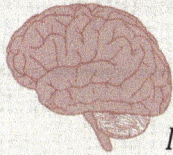

Did you know?   #42

Non-declarative memories are those that are more emotional, automatic, and procedural. The ability to ride a bike is considered a procedural memory.   We would have a hard time explaining exactly how to do it, our bodies just know. These skills are shifted from a reflective to a reflexive thought process after they are mastered (Sousa, 2008).

A Teacher's Insight

Try using 1" grid chart paper if students are struggling with drawing the shapes, especially the triangles.

Base in Balloons

Grade Level: 5-8

Materials: Inflated balloons attached to a straw or stick, such as a paint stirrer.

Objective: In the context of using movement and balloons, students will develop a deeper understanding of the base ten system.

Common Core Connection:

CCSS.Math.Practice.MP7 **Look for and make use of structure.**
CCSS.Math.Practice.MP8 **Look for and express regularity in repeated reasoning.**

Note: The common core does not specify working with bases other than base ten, however, when a deeper understanding of the base ten system is desired, then working with other bases is helpful.

Introduction	Discuss that as a class they will be exploring how the base ten system works by looking at other bases. Have 4 students sitting in chairs right next to each other. Hand the student on the far right (when facing you) one balloon, the student on his right two balloons, the student next to him, four, and the one next to him, eight.
	
	Explain that students will be working in the base of 2, which means every time there are two, a new group is formed. In other words, in each place, only 0 or 1 can exist. Start by counting from one to ten.
	  ✓ When one is called, the ones stands and raises his balloon.
	  ✓ When two is called, the student with two balloons stands and raises his.
	  ✓ When three is called, the two and the one raises his balloon,
	  ✓ Four is for the student with four balloons, and
	  ✓ Five is shown by a four and one, etc.
	The total number of balloons raised should be the number that is called.

On the board, make columns, as those in the place value chart, and mark how to represent the different numbers:

Total Balloons				
1				1
2			1	0
3			1	1
4		1	0	0
5		1	0	1
6		1	1	0
7		1	1	1
8	1	0	0	0
9	1	0	0	1
10	1	0	1	0

Discuss what makes this activity difficult or tricky to think about.

Exploration	In partnerships or small groups, students will explore base 3 and 4 (if time permits). You may choose to assign different groups a different base. Provide a variety of math tools and ask students to find a way to represent 1-10 using base 3 (remember, if there are 3, it makes a new group so there will be 0-1 on the far right column, 3-5 in the next, etc.) with concrete materials and numerically on a chart.
Discussion	Share out findings with both concrete and numerical representations, and chart on the wall. Look for patterns, like the digits that exist in the base are one less that the actual base (base 5 only has 0, 1, 2, 3, and 4, but not a 5). Compare the bases to the base ten system and explore what understand more deeply about the base ten system after exploring other bases.

Base In Balloons
Recording Chart

Base I chose_____

Total Balloons				
1				
2				
3				
4				
5				
6				
7				
8				
9				
10				

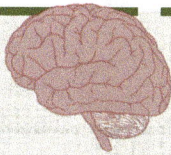

The parts of the brain that are dedicated to non-declarative memories are the frontal and basal-ganglia circuits; the parietal cortex for working memory tasks and knowledge of motor skills; the cerebellum where memory of balance, skilled movement and motor sequencing is stored; and the Broca's area which stores the process of non-motor sequences like music, timing, and rhythm (Ullman, 2005).

A Teacher's Insight

If you are frustrated because your students do not seem to be catching on the base 10 system quickly, try exploring numbers in base 2, 3, or 4 yourself. Notice the cognitive process you have to go through to make sense of the relationships. This is not much different than the 6 or 7 year old who is exploring the base 10 system for the first time.

Rock, Paper, Scissors
Is it fair?

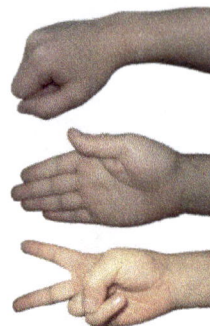

Grade Level: 7

Materials: paper, pencil

Objective: Through playing the game of *Rock, Paper, Scissors*, students will explore experimental and theoretical probabilities and how to calculate them by determining possible outcomes and use it to decide whether or not it is a fair game.

Common Core Connection:

Grade 7

CCSS.Math.Content.7.SP.C.5 Understand that the probability of a chance event is a number between 0 and 1 that expresses the likelihood of the event occurring. Larger numbers indicate greater likelihood. A probability near 0 indicates an unlikely event, a probability around 1/2 indicates an event that is neither unlikely nor likely, and a probability near 1 indicates a likely event.

CCSS.Math.Content.7.SP.C.6 Approximate the probability of a chance event by collecting data on the chance process that produces it and observing its long-run relative frequency, and predict the approximate relative frequency given the probability. *For example, when rolling a number cube 600 times, predict that a 3 or 6 would be rolled roughly 200 times, but probably not exactly 200 times.*

Introduction 1	The game will probably need no introduction, but have a couple of students come up to the front to demonstrate how to play the game, *Rock, Paper, Scissors*. Tell students that there are big competitions around the country, but you have been wondering if it is a fair game or not. Ask what the students think about it.
Exploration 1 Experimental Probability	Have students get into partnerships and play the game many times and find a way to determine if they think it is fair. Let them know that coming up with a recording strategy will help them make a case one way or the other. Also, inform them that in order to determine the likelihood of it being fair or not, it will be important to have many trials.

Discussion 1	After a period of time, when it appears that students have done enough sampling, recorded their results and come up with a theory, have the groups converge to share out what they hypothesize. Share the different models used, such as a tally chart, matrix, etc. Choose the one they should use to collect data from the entire class, post and invite the different groups to enter their data. Discuss if having more data changed their theories at all.
Exploration 2 Theoretical Probability	In small groups, students need to create some kind of model (such as a tree diagram, table or matrix) that will prove whether or not the game should be fair (theoretical probability). If students are unfamiliar with tree diagrams, it might be necessary to get them started with the hand choices from just one player. Have the students calculate the final probability of the combinations between two players from their different diagrams, and if appropriate, challenge students to create a model to show the possible outcomes for three players (they will have to decide how to determine a winner).
Discussion 2	The groups share the different models and then as a class they need to look and compare the models to see how they are the same or different. It is important to note that all models have the same information, but just represented differently. Explore how to calculate the probability for two and three player teams.

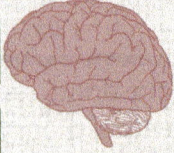

Did You know? #44

When examining the idea of neuroplasticity and how the brain can change, we saw that repetition, or rehearsal, is an important element of that change. This is especially true for moving information from our short-term to long-term memory. There is a definite chemical reaction in the brain when information goes from short to long-term memory. A protein, kinase A, "moves from the body of the neuron into its nucleus, where genes are stored. The protein turns on a gene to make a protein that alters the structure of the nerve ending, so that it grows new connections between the neurons" (Doidge, 2007, p. 220). What this means is that when we learn, we are actually changing which genes are turned on.

Make it Memorable

Allow more freedom in the way students investigate and report on material. When students have freedom of choice in what they learn and how they present it, they are more likely to learn and remember the material.

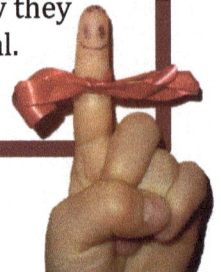

Hula Hoop
Circumference

Grade Level: 7

Materials: Hula Hoops (or large circles printed on playground), long measuring tapes, balls of yarn and scissors, yardsticks, etc., pencil, math journals or paper, sidewalk chalk

Objective: In the context of using the Hula Hoop, students will explore the relationship between diameter and circumference.

Introduction	Before going outside, or to the location where they will be using the hula hoops, ask students to tell what they know about a circle. Chart responses on the board. More than likely, students will know of some terms such as diameter, radius, and circumference. Ask student to describe each one, as well as others they might share. Ask them if they think that there might be a relationship between any of these terms. Explain that to have a relationship, one amount is dependent on the other.
Exploration	Outside, students will spread out the hula hoops in a complete circle. Allow them to determine how the radius and diameter are related and prove it using the materials provided. When they have an idea, discuss and ask how they can write that algebraically ($2r = D$). Emphasize that when finding the relationship, they found out how many of the radius are in the diameter. Write this on the playground (if possible) in chalk or have them write it in their math journals. When students are ready to explore the relationship between diameter and circumference, break the class into smaller groups and allow them to figure out how to determine how many of one is going to be in the other. Once they have an idea, students are to make a pictorial representation and label it with numbers to prove their theory. Have them find other circles, either on the blacktop or in smaller objects, and test their theory to see if the relationship they noticed works with other circles.

Discussion	Gather the groups together with the Hula Hoops and invite groups to share the algebraic equations that they created and to prove it. The groups will probably be using the term "3 and a little more" when referring to the number of diameters that can go around the circumference. Ask the class to figure out about how much that little more is. They might realize that it is about 10% if they are using string and fold it into parts. Let them know that that 3 and a little more has a special name, called pi and introduce the symbol. Now re-invite students to refine their equation with the new symbol rather than "3 and a little more."

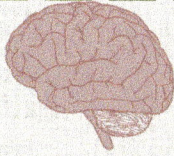

Sousa describes different types of rehearsal: *intial* and *secondary*, which involve when the rehearsal takes place; and *rote* and *elaborative*, which is the type of rehearsing being done.

Make it Memorable

Teach students how to memorize. In my experience, many student who do not know their multiplication facts simply do not study because they do not know how. They find all they have to learn too overwhelming and simply do not even attempt it. They need to be taught how to chunk information into manageable pieces and be given enough time in class to study, preferably with the teacher or a stronger student. They also need to be taught how to make and use flash cards and to separate them as they are learned.

Projects, Units & Games

Lesson Name	Grade Level	CCSSI Standard
Creating a Mosaic	Any	Synthesis of many different math skills & standards
Lines, Lines Everywhere	5	CCSS.MATH.CONTENT.4.G.A.1
Making of a May Pole	6	CCSS.Math.Content.6.RP.A.3 CCSS.Math.Content.8.G.B.7
School Survey	6	CCSS.MATH.CONTENT.6.SP.B.5
Interactive Board Movement & Math Game	Any	Fun way to practice any math standard for any grade

Creating a Mosaic

Grade Level: Any
(the above mosaic was done with a 2nd grade class)

Materials: Firing clay, tools for clay, color & clear glazes, kiln (grout & wood frame for assembly)

Objective: In the context of making a mosaic, students will learn about and apply mathematical concepts such as scale, planning, arrays, area, squares.

Common Core Connection:

This project has the potential in touching many of the math standards including measurement, computation, scale, and geometry.

Introduction	Kick off the project with a drawing contest. Give students a week to draw a picture of something that they feel is representative of their community. Have students vote on which picture should be selected for the project.

Exloration	Display the picture on a document camera (or copied onto an overhead) and display on butcher paper. Ask for the class' input on how large they think the end product should be and focus the picture on the wall to that size. In shifts, small groups can trace the projection onto the butcher paper.

The number of students in the class will determine the amount of total tiles. For example, if you have 24 students, then a 5 x 5 array might be desired. If rectangular, possibly a 6 x 4. The object is to make sure that every student has one square with which to work. Have students figure out how to divide the picture into that many squares. They will need to measure the end product and divide by the number of squares desired to figure out how long each square will be. *With younger students, this will be heavily supported by the teacher, whereas older students can work in small groups to come up with where the divisions should go.*

Once the butcher paper has the completed picture drawn and the grid drawn over it, it is important that the class decides what color each part will be painted. This is because if the students cut a flower, for example, down the middle, there needs to be an agreement as to what color that flower should be so that two students do not color each half a different color. At this time, it is now ready to be cut out and handed out to students to prepare.

Students need to measure their square, if they do not know the dimensions already, so they can make their clay tiles. Using the clay and tools, they are to roll out their clay and then cut out the square. This is tricky, especially with the younger ones. Determine an appropriate thickness (perhaps ¼ inch) for each tile so it is not too thin and to prevent the mosaic from being uneven. Have students check their thicknesses as well as the dimensions of the sides before beginning to work on the tile.

When ready, students take the picture assigned to them, and with clay tools they trace over their picture to make the same impression in the clay. Make sure that the impression has enough definition that it can be seen once dried. Have students use under glazes to paint the portion of their picture.

Fire in kiln, paint with glaze and then re-fire. When tiles are ready, mount onto a wooden frame with a plywood background and grout. |

Discussion	Since this is a lengthy project, discussion will occur throughout. The discussion focus will be determined by the age level and what concepts are being learned in class.

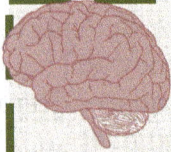

Did you know?   #46

Initial rehearsal occurs immediately when the student is presented with the information and tries to attach meaning to it. If he is unsuccessful at this moment, then the new information will most likely not be retained.

A Teacher's Insight

As written, this is a fairly ambitious project and will take some time. This same idea can be used to make a quilt or more simply with colors/paints and paper (laminated). This lesson was done with a second grade class and a similar type project of quilt-making with first graders. The level of teacher guidance will depend on the age of the students.

Lines, Lines, Everywhere!
Geometry Unit

Grade Level: 4-5th

Materials: See individual sessions below, small math journals for the unit.

Note: This is a 9-day Geometry unit that was designed for a group of 5th grade students who did not do well in the unit in their classroom using Envisions curriculum. Our school had an IXL.com account, so this was used for the practice piece for many of the sessions. The IXL lessons are included for those who may also be using this online program.

Note: IXL.com can be used without a subscription, but tracking of scores will not be possible, and the number of problems per day will be limited.

<table>
<tr><td colspan="2">Common Core Standard(s) Addressed:

Grade 4
CCSS.MATH.CONTENT.4.G.A.1
Draw points, lines, line segments, rays, angles (right, acute, obtuse), and perpendicular and parallel lines. Identify these in two-dimensional figures.</td></tr>
</table>

Session 1: Line, line segment, ray, point
Materials: markers, tape, poster paper, glue

Introduction 1	Hand out poster paper to individuals or partnerships and let them know that they will be creating a definitions poster for the unit. Every day they will add the words to their poster and write a definition and illustrate it with a picture.
Exploration/ Discussion 1	Today's vocabulary is line, line segment, ray, and point. Go through each term asking students what they understand it to mean and what picture they would draw. Help clarify any misconceptions and guide them to add them to their posters using a picture and label.
Practice/ Assessment 1	In math journals, students write a paragraph to describe each of the terms and compare/contrast them with each other. They are to include a picture as an example. Complete IXL 5th grade B.29

Session 2: parallel lines, perpendicular lines
Materials: rulers, jump ropes, cameras (optional)

Introduction 2	Introduce these two words and ask what they understand them to mean, helping to clear any misconceptions.
Exploration/ Discussion 2	Take students outside and in partnerships or groups of 3, notice around the school yard examples of parallel or perpendicular lines, record in their journals, and use math language to justify how they know. If possible, students can take pictures of their examples. Group students into groups of 4 and hand each group 2 jump ropes each. Go outside, or a space with enough room to spread out. Shout out either parallel or perpendicular. Students then take their ropes to build what was stated. Have groups justify how they know they made perpendicular or parallel lines. Challenge them by measuring the distance between the two ends of the ropes (for parallel lines) to prove that their lines are indeed parallel. On their posters, students need to figure out how to draw exactly parallel and perpendicular lines and label.
Practice/ Assessment 2	In math journals, students write a small paragraph to describe in detail the difference between parallel and perpendicular lines. Complete IXL 5th grade B.30

Session 3: finding rectangles within rectangles
Materials: sidewalk chalk or electrical tape, rectangles within rectangles handout

Introduction	Looking at a rectangle drawn on the interactive white board on grid background, have students identify what the width and height are. Have them decide how to label the quadrilateral. Add to that rectangle on the bottom and add more points and then ask where certain rectangles are, there are now three in total: ABCD, ABEF, DCEF
Exploration/ Discussion 3	Outside or in a room with a large space, draw, or create with tape, multiple rectangles which are connected together. Have students label the vertices. Then shout out points of one of the rectangles comprised in the figure and have the students run around the perimeter of the corresponding rectangle. Do this several times. Back in the classroom, students draw a picture of two adjacent rectangles, label and write the names of the rectangles found in their picture.
Practice/ Assessment 3	In their math journals, students explain how to identify one rectangle from another in a picture of adjacent rectangles. For practice, provide a worksheet that requires students to identify specified rectangles within rectangles.

Rectangles Within Rectangles

Directions: Outline the rectangles using the designated colors

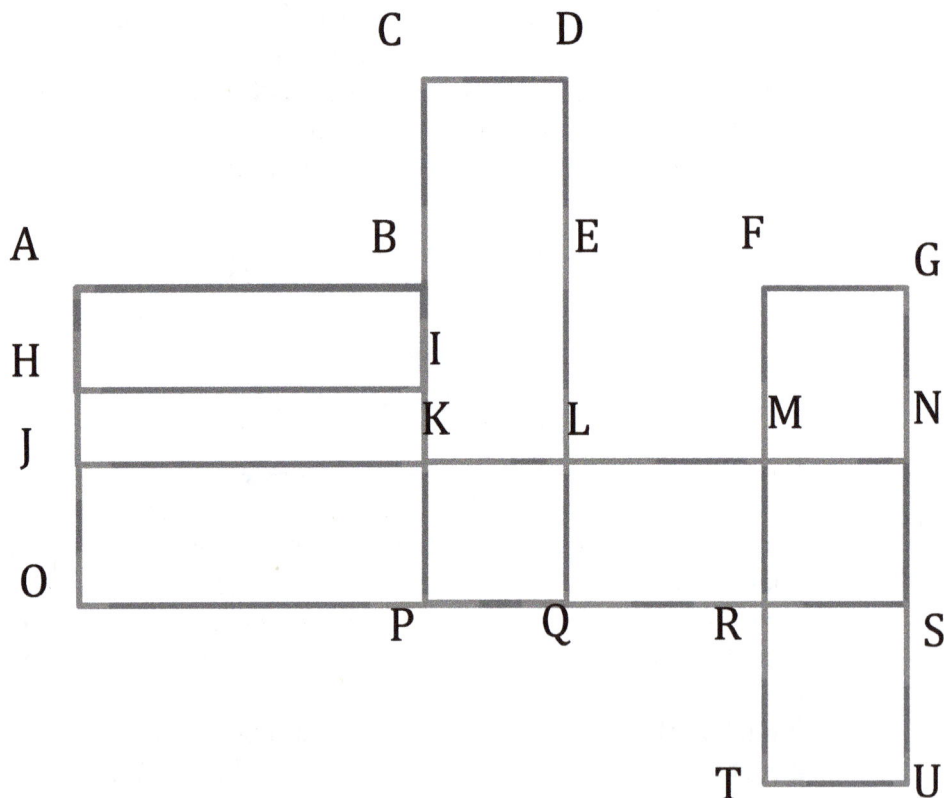

Red: ☐FGUT

Green: ☐ABPO

Blue: ☐ JNSO

Orange: ☐KMRP

Yellow: ☐CDLK

Black: ☐HIPO

Name any two adjacent
rectangles_____

Session 4: Angles & Vertices
Materials: Pipe Cleaners or straws and twist ties (insert ½ of a tie in each of two straws to join them together)

Introduction 4	Ask students to explain what an angle is by using a previous drawing (there are angles in the rectangles and perpendicular lines that they will have drawn). Show them how to appropriate label and name an angle.
	Note: Many students do not realize that the measurement of an angle is the amount of space that is in between the two lines – which is related to linear measurement in that one inch is the amount of space between the two numbers, and not just the tic marks. Students also do not realize that the point of the vertex goes in the middle when naming the angle, such as: <ABC
	A B C
	Pass out pipe cleaners or straw kits to all the students and review the three kinds of angles (acute, obtuse, and right) and have them make the angles with their pipe cleaners. Discuss the measurements of each kind.
Exploration/ Discussion 4	Go out to the playground, or around the school building with math journals to find examples of each kind of angle and record. Take pictures of the different angles and create a presentation to make into a class book or publish on your class webpage.
Practice/ Assessment 4	Add the vocabulary to the posters, label and describe with words. Complete IXL 6th grade Z.5.

Session 5: Opposite Angles are Equal
Materials: Sidewalk chalk

Introduction 5	Go outside on the playground to find an example of intersecting lines. Without giving it a name, ask what they notice about the two line segments. Students will say things like, "they are crossing". State that two lines that cross are also called "intersecting." Ask what word it sounds like (intersection) and discuss.  Ask what they notice about the angles. Students will generally say that there are 4 or that they total 360 degrees, so be more specific by marking two opposite angles with sidewalk chalk and ask what they notice about those two in specific. If they are perpendicular lines, then they will probably say that they are 90 degrees, but since all will be 90 degrees it will not suggest that opposite angles are equal. If there are no natural examples of intersecting lines creating two obtuse and two acute angles, then create some with either tape or sidewalk chalk. Now select two opposite angles and ask what they notice. They might only notice that they are both either acute or obtuse. Use your pipe cleaner to measure one and compare it to the other.
Exploration/ Discussion 5	At this point do not suggest that all angles are this way, but ask if they think this always happens or just happened this time. Lead the students around (or in a whole class, send them off in small groups) to find intersecting lines and decide if the opposite angles are equal or not. Students can record in their math journals the examples they find. Return to the class and discuss what they found out and why they think that it might be true.
Practice/ Assessment 5	Have them add the term "opposite angles" and brief description on their posters. In their journals explain in a small paragraph what they are and provide an example. *Note: the word "adjacent" might come up in this lesson and should be added to their poster if it does.*

Session 6: Supplementary
Materials: protractor, chalk, meter stick (to use as a straight edge)

Note: Session 5 & 6 may be combined onto the same day since when exploring the idea of vertical angles, adding two adjacent angles to prove that they equal 180 may be helpful.

Introduction 6	If it didn't come up in the last session, use some of the same angles that were explored with opposite angles to see what they notice with the adjacent angels. They should notice that together they make 180 degrees.
Exploration/ Discussion 6	Have someone measure one of the angles with a protractor. If using outside angles, which are painted, they would be getting approximate measurements since the lines will be quite thick. Ask them to figure out with a partner what the measure of the adjacent angle will be. Discuss possibilities and their justifications. Measure the angle to compare to what students hypothesized. As a class, generate a number sentence or equation to match the ways they figured it out (such as 180 - <A = <B). Introduce this mathematical concept as *Supplementary angles.* Using any straight line on the playground, divide it with the chalk and meter stick (to make a straight line) and measure one of the angles created. Students should test their equation with this new problem. Have students draw several intersecting lines with a straight edge in their math journals. Allow them to measure one angle and have them solve for the remaining three angles.
Practice/ Assessment 6	On their definition posters, add the vocabulary of supplementary angles and adjacent angles, draw and describe.

Session 7: Complimentary angles
Materials: protractor, chalk, meter stick

Introduction 7	Review the concept from the previous lesson of supplementary angles and how it describes how when two angles are put together they equal 180 degrees. Ask what they think would happen if they split a right angle (demonstrate on the board or on ground with chalk).

	When they say that the two together would equal 90 degrees, confirm by giving the name "complimentary angles" to this concept.
Exploration/ Discussion 7	On the playground, find a 90 degree angle, like on the 4 square court, and divide one of the angles with the chalk and meter stick. Measure one angle and have students figure out what the missing angle would be with their partner. Remember to reinforce the word "adjacent" as it will still be new for them.
Practice/ Assessment 7	On the definition posters, add the vocabulary of complimentary angles, draw and describe. In their math journal, write a small paragraph comparing and contrasting Supplementary and complimentary angles. IXL 6th grade Z.6

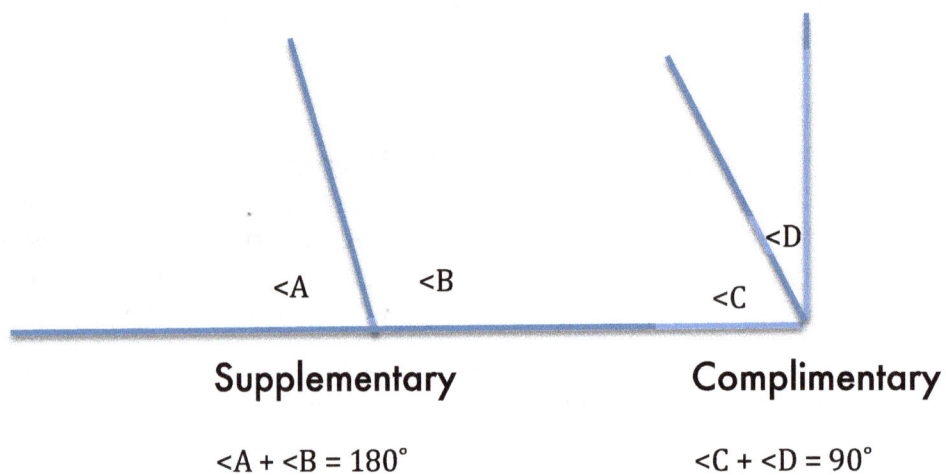

Supplementary   Complimentary

$<A + <B = 180°$ $<C + <D = 90°$

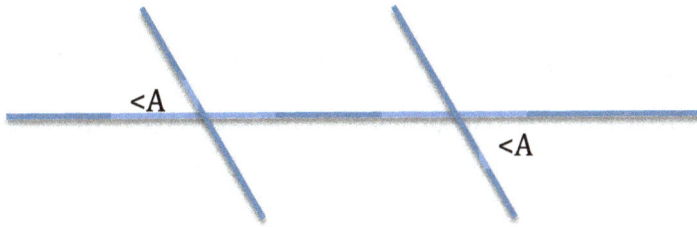

Session 8: Transversal of Parallel Lines (for 6th grade in some curricula)
Materials: Sidewalk chalk or Jump ropes, meter stick, crayons or markers

Introduction 8	Create a transversal of parallel lines on the red top and label with points. Ask students to find specific angles, line segments, etc. Ask which two angles are the same and how they know without measuring. They will recognize the vertical angles, but not the ones made by the transversal. Discuss this concept and use measurement tools, either the pipe cleaner or protractor, to prove.
Exploration/ Discussion 8	Have two students run and stand in angles of the same size. There will be 8 total angles, so 8 children should be standing at the same time. Then have students quickly swap with someone who is standing on an angle that is their same size. Have small groups make their own transversals of parallel lines and identify the equal angles, supplementary angles, and complimentary angles.
Practice/ Assessment 8	In math journals, students make a few different examples of transversal lines and mark all the angles that are equivalent with the same color. Write a small paragraph explaining how they know which angles are equal and then add to definitions posters. Complete IXL 6th grade Z.7

Session 9: Assessment

Materials: pedestrian map of Washington DC copied for each student, or on presentation board for whole class assessment.

Description	Looking at a pedestrian map of Washington DC, find an example of the following by highlighting and labeling: ✓ Acute angle ✓ Obtuse angle ✓ Right angle ✓ Line segment ✓ Parallel lines ✓ Perpendicular lines ✓ Opposite angles are equal ✓ Complimentary angles ✓ Supplementary angles On a blank sheet of paper: ✓ Draw and label an angle: ABC ✓ Draw and label a quadrilateral DEFG

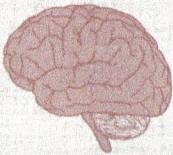

Did you know?   #47

Secondary rehearsal allows students to spend more time to make sense of the material and connect the ideas to previous knowledge, which increases the chance of retention (Sousa, 2008).

Make it Memorable

As teachers, we need to slow down when we are presenting new material to make sure that there is comprehensible input, especially if there are second language learners in the class, and offer more opportunities for initial rehearsal, and to provide longer and more frequent breaks between information to allow students to individually and collectively reflect on what is being presented and engage in secondary rehearsal. If we reflect on the retention time period illustration, it might be wise to allow students to engage in secondary rehearsal after no more than 20 minutes of lecture.

Making of a May Pole
Ratios & Proportions

Grade Level: 6

Materials: Tether ball pole, Measuring tape, ruler or yardstick, stick – such as a 24" dowel, paper towel insert (optional), grid paper, several spools of 2-3" wide ribbon of different colors, scissors, eyelets (one for every strand of ribbon used), one large carabiner spring snap hook, calculator

Common Core Standard(s) Addressed:

Grade 6
CCSS.Math.Content.6.RP.A.3 Use ratio and rate reasoning to solve real-world and mathematical problems, e.g., by reasoning about tables of equivalent ratios, tape diagrams, double number line diagrams, or equations.

Grade 8
CCSS.Math.Content.8.G.B.7 Apply the Pythagorean Theorem to determine unknown side lengths in right triangles in real-world and mathematical problems in two and three dimensions.

Objective: In the context of creating a May Pole, students will practice/apply using ratios to calculate unknown sides of triangles. Students will also be introduced to the Pythagorean Theorem to calculate the hypotenuse of a triangle.

*Note: Students will need to have already learned how to solve for unknown sides of similar triangles or other polygons

1 - Finding the Length of the Pole / Heights & Shadows

Introduction	Introduce to the students the concept and history of a May Pole. There are you-tube clips that show students dancing with a May Pole. Let students know that they will be creating a May Pole that they, and possibly other students, can use. It will be more fun if there are multiple tether-ball poles to use. Explain to the students that in order to create the May Pole, they will need to know how tall the pole is so that they can calculate how much ribbon they are going to need. Inform them that they are going to use shadows to calculate the height.

Exploration

In teams, students go outside on a sunny day at a time where there is a visible shadow of the tether-ball pole. They will also need a shorter object that can cast a shadow (ex: dowel or ruler). Students measure the shadow of the pole and record on grid paper (one square for each inch). They will then hold their smaller pole and record its measurement on a grid paper as well (in pictures and numbers).

Ex:

24   3

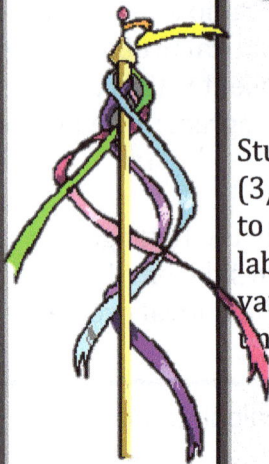

Students are to set up calculate the ratio between the two shadows (3/24 = 1/8 = .125; or 24/3 = 8 from above example) Then they are to draw the height of the smaller object with known height and label. They will sketch the height of the pole and label it with a variable. Have them apply the ratio from the shadows to find the unknown height of the pole.

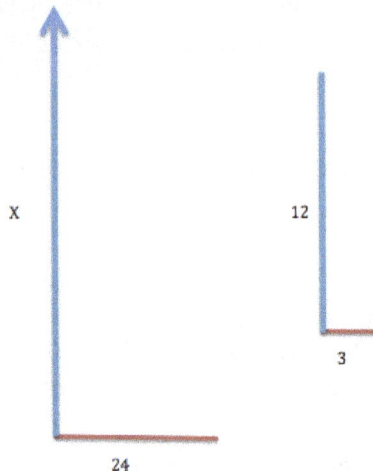

X

12

24

3

Method 1:
If $24/3 = 8$
Then $X/12 = 8$
$X = 8 * 12$
$X = 96$

Method 2:
$$\frac{24}{3} = \frac{X}{12}$$

$24*12 = 3x$
$288 = 3x$
$288 \div 3 = x$
$96 = x$

96 inches = 8 feet = height of the pole

Discussion	It is likely that teams of students will not get the same measurements as each other. If measuring the same pole, it will be due to the movement of the sun, which affects the lengths of the shadows. Discuss the different variables that might have caused students to get different values. Using a step-ladder, confirm student answers by actually measuring the pole with a tape measure.

2 - Finding the Length of the Strands

Introduction	Now that students know how tall the pole is, they need to figure out what is a reasonable size strand. First ask them to figure out how long the ribbon should be if the directions you found online suggested it to be two times the pole height. Discuss results and strategies used.
Exploration	Now tell them that online directions actually suggest 1 ½ times the pole height. In partnerships or small groups, have them figure out how long that would be. Tell the students that there is a formula that can tell how long the strand would be from the tip of the pole to the edge of the circle (or spot where students plan to stand). First go outside and have students measure the distance from the pole to the edge of the circle (for our pole this distance was 10 feet). Have students draw a sketch of the pole, distance to the circle, and ribbon. Ask students which part of this drawing would be considered the base and the height? Inform them, if they do not already know, that the line for the ribbon is the hypotenuse. Have them label the base A, the height B, and the

	hypotenuse C. Then tell them that to find C the formula $A^2 + B^2 = C^2$. In small groups students try to figure out how to get C, supporting as needed depending how much experience they have with exponents and square roots at this time. Calculators will be necessary for this. they should use have your
Discussion	Discuss if their amount was the same as what they got when they calculated 1 ½ the height of the pole. Go outside with the class. Using the ribbon, mark off the amount the students suggest from both calculations. Roll out that much ribbon and discuss if that seems adequate or if they should use a lesser or greater amount. Note: With ours, we added an extra 3 feet so we were not holding on to the very edge of the ribbon.
Next Steps	Have students cut strands of ribbons to the desired lengths. As a class decide how many strands are needed. When complete, add large eyelets to the tip of each one, then connect them with the carabiner snap hook and attach to the loop at the top of the tether-ball pole. Have students watch video clips from the internet to learn how to dance to create the braid around their May Pole.

*Before dancing the May Pole, share Video clips found on the internet that show how the May pole dance is commonly performed. Although the origins are from northern Europe, the Japanese have done a wonderful, and perhaps much more intricate, performance with the May pole.

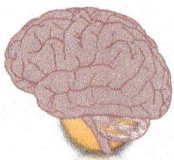

When you practice something, it gets easier for the signals to cross the synapse. That's because the contact area becomes wider and more neuro-transmitters are stored there.

A Teacher's Insight

I made the mistake of trying to explain how to do the May pole dance while at recess when dozens of first graders wanted to join in. Since we needed extra people, I let them in, but needless to say, it was a jumbled mess. I learned that it is much more effective to first share Video clips found on the internet that show how the May pole dance is commonly performed. Although the origins are from northern Europe, the Japanese have done a wonderful, and perhaps much more intricate, performance with the May pole.

School Survey

Grade Level: 4th-6th

Materials: Graphs and statistics from newspaper or online sources.

Objective: In the context of conducting a school survey, students will consider a question that is worthy of finding out data to answer, decide how the data should be collected, and evaluate the results using procedures that are in the standards of their grade.

Common Core Standard(s) Addressed:

Grade 6
CCSS.MATH.CONTENT.6.SP.B.5
Summarize numerical data sets in relation to their context

Note: The amount of teacher involvement will depend on if the class has previous experience with different data gathering and evaluation concepts. The teacher may choose to use this projects as the vehicle in which to address the individual concepts, thus making the process longer.

Data and statistics have become an important component to the common core and many problem solving skills are involved when conducting and analyzing data. After the initial project, you may wish to make this type of data collection and evaluation a regular part of your classroom experience.

Description	Announce that the school newspaper, student council report, or bulletin would like to include information or opinions of the students from the school. They have commissioned their class to help with the data gathering. Go to newspapers or online sources and explore surveys and what types of questions are being asked. Have small groups generate a few questions that they think the administration and student body would want to know. As a class, discuss the options and rank them in order of interest or importance. If the questions generated are too superficial, discuss and have them try again.

When a question has been identified, ask the class how they think they should best collect the necessary data. This includes the instrument they should use, how and when to ask the other students, what ages should be included in the survey, etc.

Class should split the jobs for conducting the survey. For example, one group can make the instrument, another group can man stations at a central location (or different locations) to collect data, and another group can compile the data.

Questions that the class needs to decide are: whether they will go from class to class or to have a stand at recess, if they want the input from the younger students, or just older, and if they want a random sample or to survey everyone.

Once the data has been collected, they need to decide how to best represent it and interpret the results. One group may want to be in charge of writing a short report describing the question, results, and pose a possible interpretation.

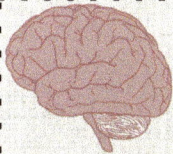

Did you know?   #49

Rote rehearsal is used when trying to memorize, for instance the multiplication tables or other math facts, but *elaborative rehearsal* is used to associate and connect the new information to prior knowledge and looking for relationships and patterns.

Make it Memorable

If we understand the multiple representation model that is illustrated by Lesh, Post, and Behr (1987), then we can help students appropriately engage in elaborative rehearsal.

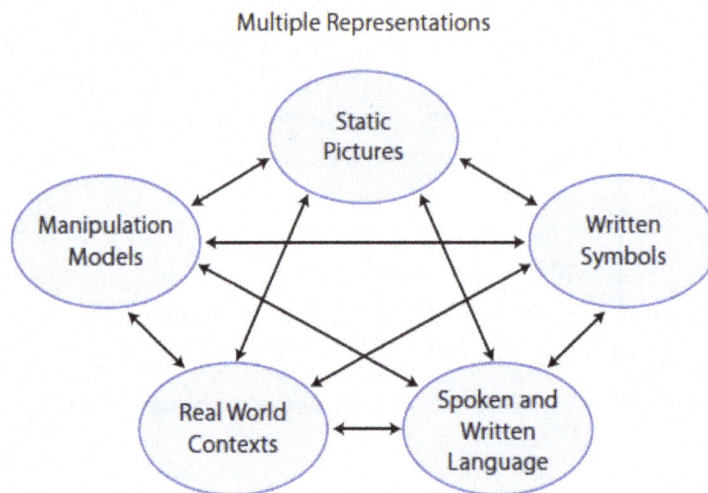

Multiple Representations

Model of Multiple Representations in Mathematics, Lesh, Post, and Behr (1987)

This model illustrates the different ways and modalities in which students should engage with the mathematics. Students may be working with written symbols when using rote rehearsal to learn their facts, but elaborative rehearsal requires that they manipulate the mathematics and make connections.

Interactive Board Movement & Math Game

Interactive white boards, is a great way to get your students up and moving. On SMART Exchange at http://exchange.smarttech.com/ , the online resource for lessons and activities, there are many templates that can be used to get children up and out of their seats during math time.

My favorite is the *Koosh ball template*, which is a free download. Each color circle is connected to its own page. Students take a Koosh ball, or something similar, throw it at the board, and their task will appear.

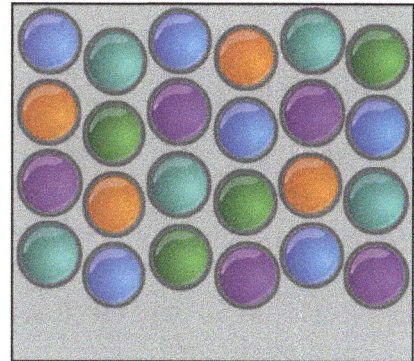

I have used this to create a Movement Game. About 50% of the tasks are movement related and 50% are math related. It would be best to teach students the movements and take pictures of students doing an exercise correctly and using that picture on the respective slide.

Grade Level: Any

Common Core Standard(s) Addressed: Any

Materials: Interactive White board, supporting materials depending on questions that are selected.

Objective: In the context of playing the Movement Game, students will practice math skills and problem solving tasks as well as engage in a variety of exercises.

Example Movement Activities ↓	Example Math Questions ↓
Hold an Elbow Puller stretch for 10 seconds (each arm)	What is the difference between area and perimeter?
Perform "supermans" for 30 seconds	Order these fractions from least to greatest: ½ 1/9 1/3 1/5
Jog in place for 1 minutes	Find the Missing Number: | x	12	26	30	36
---	---	---	---	---
y	8	22	26	?	
Perform 30 arm circles, forward and backward.	For the bake sale, week one 866 cookies were sold and week two 154 cookies were sold. About how many cookies were sold altogether?
Hold a "hurdler" stretch for 10 seconds each leg.	Round to the nearest million: 143,479,806
Squat and jump tall - do while counting by 2's starting with 3.	True or false? $15 = 3 \times 5$
Do 10 push-ups	What is a prime number?
Hold a butterfly stretch for 10 seconds	Last Saturday Matthew and Diego rode their bikes. Matthew rode 41 miles and Diego rode 27 miles farther than Matthew. How many miles did they ride altogether?
Do 20 crab kicks	A crate of apples contains 2 boxes. There are 2 layers of apples in each box. There are 4 apples in each layer. How many apples are in a crate?
Hold a plank for 30 seconds	Draw perpendicular lines. Describe how you know that they are perpendicular.
Perform 20 curl-ups	Who jump roped faster? A: 10 jumps in 20 seconds B: 15 jumps in 30 seconds
Perform 30 jumping jacks	Solve:

	22 - (14 + 7) x 2
Do 10 sit-ups	Find the area and perimeter of the following shape: 2 4 6 10
Stand on 1 foot for 1 minute while counting by 3's	Write this number: four and five hundred sixty-two thousandths
Do 5 sets of lunges.	What is the least common multiple of 6 and 7?
Do 10 squats	What is the smallest number you can make with these numbers? 3, 5, 2, 1, 9
Do 20 cross crawls	What fraction, decimal and percent does this represent?
Do 5 burpees	How many lines of symmetry does the following object have?
Draw a large infinity sign in front of you 5 times with each arm and 5 times with both arms	What is the sum, difference, product, and quotient of the following numbers: 6 & 3.

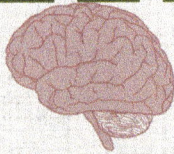

Anxiety floods your body with adrenaline, which makes it hard for neuro-transmitters to carry messages across the synapses in your brain. This in turn causes "blanking out" on a test.

A Teacher's Insight

This is a class favorite. Make sure that the math activities are updated to reflect what you are currently teaching / reviewing.

Part 3:

Getting Started With Math & Movement Stations

Movement refines and speeds up not only proprioceptive processing, but also auditory processing, visual processing and the integration of all three to produce balanced, calmer individuals who are more proficient in reading, spelling, math and writing.

--Barbara Pheloung

Brief Snapshot of our Senses

> *Every movement is a sensory-motor event.*

The following are brief descriptions of our senses. For more detailed information on each sense and how it relates to learning, as well as movement activities to target each sense, please refer to the resources and website at the end of this section.

Vestibular Sense

The vestibular system is considered the entryway to the brain and is said to have the most important influence for everyday functioning. It is "the unifying system that directly or indirectly influences nearly everything we do," (Hannaford, 1995, p. 38).

The vestibular system is located behind the mastoid bone (the lump behind the ear lobe) and part of the inner ear. It makes up part of the vestibulocochlear system, and takes the form of three semicircular canals, which are filled with fluid and tiny hair cells. Each time we move our heads, the fluid moves and bends the hair cells, which in turn sends sensory nerve impulses to the brain, particularly the cerebellum, which monitors and makes adjustments in the muscle activities, including eye movements. It allows for our muscles to adjust instantly so we don't lose our balance. The reason we tend to get more dizzy as we age is because as we grow older, reproductive hormones causes the liquid to thicken and the hair cells to be bent longer, thus taking more time to return to a comfortable equilibrium.

Tactile and Proprioceptive Sense

The sense of touch is perceived by movement of air or pressure over the skin with receptors in the dermis, which detect deep pressure, vibration, light touch and temperature. "Touch right after birth stimulates growth of the body's sensory nerve endings involved in motor movements, spatial orientation and visual perception. Touch is a strong anchor in behavior and learning. Whenever touch is combined with other senses, much more of the brain is activated and more complex nerve networks are built" (The Brain-Movement Connection, n.d., para. 14). "The touch of closely bonded family members increases activity in the hippocampus, an important center for spatial and general learning and memory" (Hannaford, 1995, p. 46).

Proprioception, on the other hand, is the sensation from muscles, tendons and the vestibular system, and is the body's sense of itself in space. The proprioceptive receptors are located in all of the muscles and they sense the degree of stretch in the muscle. Proprioception helps with vision by receiving constant feedback to adjust the shoulder and neck muscles, they keeping our eyes steady as we are reading and moving.

Vision

Whereas the vestibular, tactile, and proprioception sense aide in neurological organization, vision and the other commonly known senses are the output of this organization. "When the body and head move, the vestibular system is activated and eye muscles get stronger. The more the eyes move, the stronger the muscles of both eyes get so they can work together. When the two eyes can move together efficiently it will be easier for a student to focus and track words on a page," (The Brain-Movement Connection, n.d., para. 15).

> *Vision is the end result of neurological organization.*
> *-Svea Gold*

Bilateral Integration

Bilateral Integration is not necessarily one of the senses, but it is worth noting in this section since it fundamental to the learning process. The corpus callosum, a thick band of fibers, is a structure of the brain that is involved with connecting and integrating its two halves, with approximately 250 million nerve fibers (Lengel & Kuczala, 2010) . The more that the two sides work together, the more connections are made between the halves and the denser this area becomes. In fact, Researchers have found during autopsies that the corpus callosum is much denser in those who had professions, such as musicians and singers, whereas in individuals with dyslexia it is significantly smaller than in average individuals (Hannaford, 1995).

Cross lateral movements, like crawling, activate large areas of both hemispheres in a balanced way. It "makes it possible for nerve networks to form and myelinate in the corpus callosum. This makes communication between the two hemispheres faster and more integrated for high-level reasoning, (The Brain-Movement Connection, n.d., para. 23

> *A well lateralized brain is a more organized brain.*
>
> *-Sally Goddard Blythe*

Movement Stations

Setting up movement stations in your own classroom or an empty room (that can be shared with other classes) is an ideal way to allow children to take a break, move, stimulate all necessary sensory systems while at the same time allowing the brain to practice and process math skills. This will allow students to access and use the entire brain while engaging in math exercises. *Note: reading skills can also be substituted for math when creating movement stations.*

There are a few commercial and non-profit programs, such as the Minnesota Learning Resource Center, and Ready Bodies Learning Minds, that encourage the use movement stations as a way to enhance brain and motor development. The testimonials found on their websites, from occupational therapists, speech pathologists, physical therapists, parents, and others, attest to huge gains seen in children after 6-9 months.

A couple of years ago I became very serious about embedding the movement stations into my math program. I assessed all my students in both their developmental math level as well as their motor skills, both at the beginning and at the end of the year. I found that there seemed to be a correlation between the gains in motor skills and increase in developmental level in mathematics. Some of the students were being monitored by the school's student study team and were able to be discontinued by the end of the year. Some students made dramatic increases, many others moderate, and to be honest, a couple made very little progress at all. I believe that the movements not only addressed their physiological needs, such as an impaired vestibular system and many un-integrated primitive reflexes, but it also allowed the students to be more open to learning.

An empty classroom can be easily converted into a movement room, which not only focuses on movements that encourage sensory integration and motor development, but that incorporates practice of math skills as well. If an entire room cannot be spared, stations can still be set up within a classroom, outside or perhaps a shared area such as a loft, hallway or cafeteria. Most of the materials probably already reside within your school walls, many in the PE room.

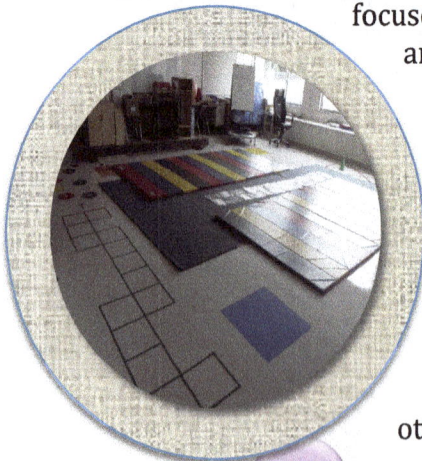

On the following pages is an example list of stations and directions of use. Some only target movement and others only target math skills, but most integrate the two.

Set up as many stations as your space allows, whether just 4 or 15 stations. The key is that the stations vary in purpose. Try to have at least one which targets the vestibular system, tactile & proprioception, visual system, and bilateral integration. Modify the math activities to meet the needs of your particular students and be sure to update them periodically.

Equipment used in the described stations are: 12 foot balance beam (single board can be used too), polyspots, small wooden stepping stool, two large gym mats, creeptrack (made with clear plastic shower curtain and transparent pockets), rebounder, pocket charts, sentence strips, white board, boxes, clothespins, solid small ¾" wooden cubes, sticky Velcro, and electrical tape. With the exception of a few items, such as the mats and balance beams, most of the materials are commonly found in a classroom or easily purchased at a local store.

The stations are organized as a circuit and students rotate through them as many times as time permits. The goal is not to be fast, but to do the activities with precision and for students to focus on their bodies. When they are stuck behind one another, they must do a movement like cross crawls (marching and touching their knee with their opposite hand) while skip counting as they are waiting, so their brain is continually working.

When treated like a circuit, where students move from one to the other until

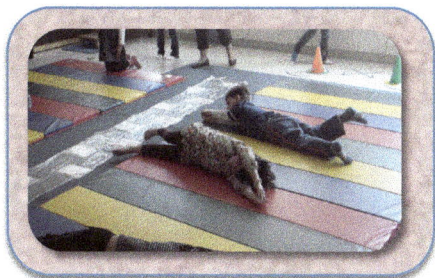

complete, movement stations tend to work best with a small group of students (up to about 6). However, entire classes may also participate, but rather than rotate through all the stations, it has proven to be more manageable if the stations are grouped into two or three and a set amount of students (like 5) move through those two stations over and over in a two minute period. After the two minutes, the groups rotate. A total of 10 stations, for example might be broken into five groups. At two minutes per group, that would be approximately 10 minutes for one full rotation.

If you choose to establish stations within your classroom, then you could have it run during centers time with one small group on the circuit, or simply when you see a student who needs a mental break, send him to one station and have him re-join the group when ready. Even setting up one corner of your classroom as a "relax and reorganize" corner where there may be a menu of movements, trampoline, rhythm box, rocking or spinning chair, etc., available for a child who needs a chance to reorganize and regroup before continuing with a whole-group activity.

"When large-motor movement precede small-motor movements, the small-motor control will then be freer and less constricted" (Dennison, 2006, p.66). This means, that children with fine-motor challenges need to work with their large gross motor muscles first. When engaged in a movement, if the student shows signs of stress or becomes over-excited, then it is necessary to reduce the type and/or amount of stimulation.

Any movement program is most effective when it is done on a daily basis for at least 3 months (Goddard Blythe, 2007). If exercises are stopped, for whatever reason, and the

child starts to regress, it will be important to return to them. This means that the connections are not yet permanent. Movement helps the myelin sheath grow around the nerve cell to protect it and allow it to fire rapidly, therefore sometimes a high fever will jeopardize a newly connected nerve cell that has not been myelinated enough.

Especially if language is an issue, it is critical that we get children to verbalize as much as they can while doing movements (Pheloung & King, 1993). At first this may need to be modeled by the teacher, just as any new skill. This not only heightens awareness of with directionality are lacking positional understanding of their own body. If they have not explored enough in their 3-dimensional space around them, they will also struggle with visual spatial skills, which is an important component in mathematics.

An example of using language:
- *Teacher* - "What hand are you using? Which part of your body is starting to roll? Tell me everything you are doing as you are doing it."
- *Student* - My right foot is forward and I am pointing to it with the first finger of my left hand" (Pheloung, 1997).

Examples of general body awareness activities also supporting language development (from Pheloung & Liljeqvist, 2004, and Pheloung, 1997, and Pheloung & King, 1993) are:
- Letting the child walk anywhere around the room until you call "stop!" and tell him what to do "melt like ice cream...walk...Wobble like jello...walk...stop rigid like a telephone pole."
- Crawl and roll blindfolded. Have the child describe his movements, what he feels, etc.

Giving the Brain a Workout

The following pages is an example of a circuit that combines movement with math activities

1- Hopscotch

Students hop on the hopscotch using the foot that is on the same side of the square. They should jump with both feet on the double squares. Numbers or expressions can be taped to each square and students have to read/solve each as they hop. Math vocabulary can be used as well.

2- Color Circles

Using both feet together, students jump on the circles saying the color where they land. If older, simple equations or shapes can be taped to each circle and students have to state the answer with each jump.

3- Balance Beam

Students walk on the balance beam (can even use a 10-12 foot piece of lumber on the ground) slowly putting one foot exactly in front of the other (toes touching heel). If students can easily balance while looking forward, they can try touching their nose with their finger, alternating hands and keeping the opposite outstretched. They can also balance a beanbag on their head or look up while walking. Challenge them to go backwards as well or skip count wile walking.

5- Rhythm Box

Perfect activity to practice skip-counting.

Prepare sentence strips, each with multiples of a number (2, 3, 4, 6, 7, 8, 9 – I do not include 5's or 10's because my students already know those) and attach to the wall with velcro. Write the numbers from bottom to top in alternating colors red and black. Students step up and down the rhythm box. Example: When skip counting 2's, they say "2" (as they step up with right), "4" (as they step up with the left), then do not say anything while stepping down. "6" (right), "8" (left), and so on. Do not allow them to go too fast.

6-Spinning chair

Students can either spin themselves or best to have an adult spin so the chair doesn't move all over the place. Make sure to err on the side of spinning too slow unless you are certain the student can handle going faster. Spin in both directions about 15 seconds or less. **Do not include a spinning station right before one that requires balance. Stop Immediately when dizzy. Spinning can cause some children to get sick or a headache**

7 – Infinity Walk

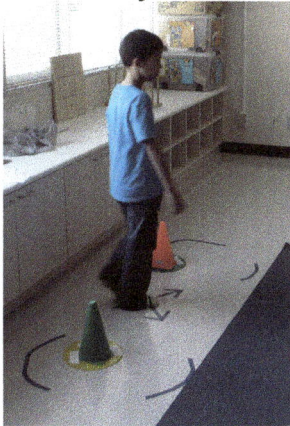

Either create a large infinity sign with electrical or painters tape on the floor, or put two cones, stools, etc. about 4 feet from each other. Find a spot on the wall that is perpendicular to the center of the infinity sign (where it crosses) and mark it with an X that is easily visible to the children and at eye-level.

Have children walk on the tape, or around the cones in a pattern of the infinity sign keeping an eye on the X the entire time. When going around, the trick is not turning their body, but just turning their head quickly to keep their eyes on the X. For very young children (4 and 5), they can just follow the path without looking at the X on the wall until they are used to it.

8 – Fine Motor eye/hand coordination Station

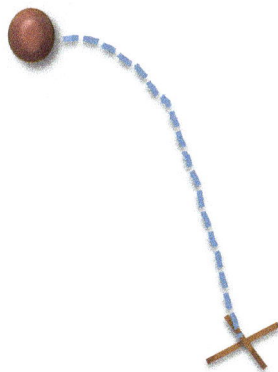

Choose 1:

Ring toss – Using a board with dowels sticking up, children throw rings onto the dowels

Clips on box – Provide a plastic or cardboard box with clothes pins that are clipped around the edge of the box. Students unclip the clothes pins from the box with their non-dominant hand and place them inside. If pins are already off, have them put them (or some of them) on the box with their non-dominant hand. The stiffer the clothespin, the better.

Ball bounce: Mark a spot on the floor with an x or sticker and have the child bounce a small rubber ball to hit exactly the marked spot.

215

9-Ordering Numbers

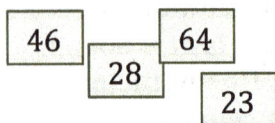

46		64
	28	
		23

Provide 4-6 index cards with numbers written on them and have students order them from either least to greatest or greatest to least. Include number types (whole numbers, decimals, fractions, integers) appropriate for your class.

Note: provide a variety of sets of varying degrees of difficulty. If a set is too hard, students will get stuck in this one station. Some Options:

9- board work

Trace the infinity sign – about 2 or 3 times with each hand. Stand in the center with nose at the level where the sides cross.

Rainbows- Children draw large rainbows that start from one side of the body and end on the other side. Point is to get large motion with arm and cross midline.

Tornadoes – Children start out and draw large circles that cross the midline and draw them in the form of a spiral that ends at the center of their nose.

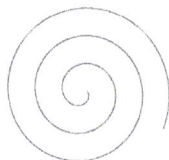

Mirror Writing – use two hands and write or draw. The two pictures should look like a mirror image of each other.

10- pencil rolls

Children lay on the end panel of the mat with their arms up over their head along each ear and roll along the length of the mat. Watch for slow moves and straight legs. Their eyes can be either open or closed (might want to encourage open the first round and closed the second).

You can also have a string of numbers or number sentences posted at eye level at one end of the mat that students are to say/solve each time they roll. This helps with vision as well.

Do not include a rolling station right before one that requires balance. Have students stop Immediately when dizzy. Rolling can cause some children to get sick or a headache

11- Creep Track

Using sight words/numbers (2-3) that students already know (not unfamiliar words), place them in a random order in the pockets of the creep track. Have students creep (crawl on hands and knees) and say the word as their hand slaps down on it. Make sure they are going slow and are saying the word at the same time as slapping.

Use math words like: sum, difference, product, quotient, quadrilateral, parallelogram, area, perimeter, or any other that you want your students to really internalize.

12-building 3-D models (visual-spatial skills)

Provide a bucket of solid block cubes and several pictures of arrangements (from easy to more difficult). Students choose a picture, estimate how many cubes it will take to build and then build the figure shown.

13 – army crawls (aka: lizard crawls, alligator crawls)

Students are to lay on their stomachs with one arm up farther than the other, palms flat on the mat. On the side with the outstretched arm, the leg should be straight. The other leg should be bent with toes dug into the mat.

Students should push off with their toes from their bent leg and their outstretched (opposite) hand should be pulling them. Their hands and legs should switch so the opposite is outstretched.

This is probably the hardest station because children have trouble using their feet to help propel them on the mat. They will tend to drag their legs by using their hands. They also might start to put arms under their chest or lift their chest off the ground. Encourage to dig their toes into the mat and to push off with their feet. Encourage them to go slow and not to race down the mat.

14-Open Number Line

Post a sentence strip on the wall or bulletin board with Velcro.

1-100: On the front side of the strip label a "1" and "100" on the ends and on the back write 5, 10, 15, 20, etc. Write the 5's smaller than the tens.

0-1: On the front side of the strip label a "0" and "1" on either end. On the back of the strip, label as many fractions and decimals as appropriate.

Add a clothespin to the strip. In a plastic sleeve, display a number you wish the students to estimate. When they reach this station, students estimate where the number is on their number line. You may either allow students to check after guessing, by flipping over the strip, or have them wait for you to check.

Having a metronome playing at 54-60 beats per minute in the background is a great way to help students internalize the ideal learning tempo. It will also help the students from going too fast. Metronome apps can be downloaded for free on ipods or iphones and plugged into speakers.

For a more comprehensive list of movements and activities that support the vestibular system, tactile and proprioceptive sense, vision, bilateral integration, rhythm and timing and the integration of primitive reflexes, please refer to my website at http://wholechildlearningsolutions.com or any of the resources listed on the following page.

Additional Resources

These are some of the resources, books and programs, which I have found most helpful in developing and refining my understanding of movement, and brain neuroplasticity. These will be useful if you are interested in learning how to support your students through movement.

- ✓ Braingym.org
- ✓ Eyecanlearn.com
- ✓ *If Children Came with Instruction Sheets* by Svea Gold
- ✓ INPP and books by Dr. Sally Goddard Blythe, director
    - o *Attention, Balance and Coordination: the A.B.C. of Learning Success*
    - o *Reflex, Learning and Behavior: A Window into the Child's Mind*
    - o *The Well Balanced Child: Movement and Early Learning*
- ✓ *Integration of Infant Dynamic and Postural Reflex Patterns – MRNI: Neurosensorimotor Reflex Integration:* by Svetlana Masgutova
- ✓ Interactive Metronome™
- ✓ Movements that Heal by Dr. Harold Blomberg: blombergrmt.com
- ✓ Move to Learn Programme: Barbara Pheloung from Australia
- ✓ Ready body Learning Minds (www.readybodies.com)
- ✓ SMART movement program: A Chance to Grow (http://actg.org) from the Minnesota Learning Resource Center
- ✓ *The Brain that Changes Itself:* Book by Norman Doidge
- ✓ *The Woman Who Changed her Brain*: book by Barbara Arrowsmith-Young
- ✓ *Wired to Learn, Integrated Learning Therapy*: video and guide by Shirley J. Kokot

The website I have created which pulls these and other resources together for parents and teachers is: http://www.wolechildsolutions.com .

I also have developed a developmental math assessment called *A Qualitative and Quantitative Developmental Math Assessment and Intervention Protocol.* This is an individualized assessment and is accompanied with an intervention protocol for children who are struggling in grades 1-6. This assessment guide comes with fillable worksheets for easy tracking of student data. Guide can be found at www.developmentalmathassessment.com .

Appendix A:
Common Core Overview

The following charts outline the broad math standards in grades 1-6. It should be noted that some standards are found in multiple grades.

Counting	
K	Know numbers and count sequence
K	Count to tell number of objects
K	Compare numbers
Operations and Algebraic Thinking	
K	Understanding addition as putting together and adding and understanding subtraction as taking apart and taking from
1/2	Represent and solve problems involving addition and subtraction
1	Understand and apply properties of operations and the relationship between addition and subtraction
1/2	Add and subtract within 20
1	Work with addition and subtraction equations
2	Work with equal groups of objects to gain foundations for multiplication
3	Represent and solve problems involving multiplication and division
3	Multiply and divide within 100
3/4	Solve problems involving the four operations and identify and explain patterns in arithmetic
5	Analyze patterns and relationships
4	Gain familiarity with factors and multiples
5	Write and interpret numerical expressions
5	Use equivalent fractions as a strategy to add and subtract fractions
5	Apply and extend previous understandings of multiplication and division to multiply and divide fractions

Measurement and Data

K	Describe and compare measureable attributes
K	Classify objects and count number of objects in categories (related to counting)
1	Measure lengths indirectly and by iterating length units
1	Tell and write time
1	Represent and interpret data
2	Measure and estimate lengths in standard units
2	Relate addition and subtraction to length
2	Work with time and money
2/3/4/5	Represent and interpret data
3	Solve problems involving measurement and estimation of intervals of time, liquid volumes, and masses of objects
3	Geometric measurement: area – relate to multiplication and addition
3	Geometric measurement: perimeter (recognize as attribute and distinguish from area)
4	Solve measurement problems and convert large to small unit
4	Geometric measurement: understand concepts of angles and measure angles
5	Convert like measurement units within a giving measurement system
5	Geometric measurement: understand concept of volume and relate volume to multiplication and addition
6	Develop understanding of statistical variability
6	Summarize and describe distributions

Number and Operations: Base 10	
1	Extend the counting sequence
1/2/5	Understand place value
1/2	Use place value understanding and properties of operations to add / subtract
3 /4	Use place value understanding and properties of operations to perform multi-digit arithmetic
4	Generalize place value understanding for multi-digit whole numbers
5	Perform operations with multi-digit whole numbers and with decimals to hundredths
Ratio and Proportion Relationships / The number system	
6	Understand ratio concepts and use ratio reasoning to solve problems
6	Apply and extend previous understandings of multiplication and division to divide fractions by fractions
6	Compute fluently with multi-digit numbers and find common factors and multiples
6	Apply and extend previous understandings of numbers to the system of rational numbers.

Geometry	
K	Identify and describe shapes
K	Analyze, compare, create, compose shapes
1/2/3	Reason with shapes and their attributes
4	Draw and identify lines and angles and classify shapes by properties of their lines and angles
5	Graph points on the coordinate plane to solve real-world and mathematical problems
5	Classify 2D figures into categories based on their properties
6	Solve real-world and mathematical problems involving area, surface area and volume

Common Standards at a Glance

	K	1st	2nd	3rd	4th	5th	6th
Counting and Cardinality	X						
Operations and Algebraic Thinking	X	X	X	X	X	X	
Number and Operations – base 10	X	X	X	X	X		
Number and Operations with fractions							X
Ratio and Proportional Relationships							X
The Number System							X
Measurement and Data	X	X	X	X	X	X	
Statistics and Probability							X
Geometry	X	X	X	X	X	X	X

Appendix B:

The following chart outlines the Common Core standards by grade level and matches them with the lessons from this book.

Standard	Grade	Unit / tool used	Lesson Name	Page
Mathematical Practices				
Note: All lessons contain elements of the 8 mathematical practices. Lessons listed here are ones that do not address a specific content standard.				
CCSS.Math.Practice.MP4	1-2	Giant 100 grid lessons	Connecting the number line to the hundred chart	30
	3-6	Jump ropes	Human Pie Chart	129
	Any	Odds & Ends	Venn Diagrams	142
			Get the Picture?	147
CCSS.Math.Practice.MP5	1-3	Giant 100 grid lessons	Building the 100 Chart	23
	Any	Odds & Ends	Venn Diagrams	142
CCSS.Math.Practice.MP7	1-3	Giant 100 grid lessons	Building the 100 chart	23
	Any	Odds & Ends	Get the Picture?	147
	5-8		Base in Balloons	165
CCSS.Math.Practice.MP8	5-8	Odds & Ends	Base in Balloons	165
Kindergarten				
CCSS.Math.Content.K.CC.A.2	K	Yardsticks, Rulers, Tapes	Human Number Line	61
1st Grade				
CCSS.Math.Content.1.OA.C.5	1	Yardsticks, Rulers, Tapes	Human Number Line	61
CCSS.Math.Content.1.NBT.C.5	1-2	Giant 100 grid lessons	Add 10	25
CCSS.Math.Content.1.G.A.2	1	Odds & Ends	Body Shapes	151
CCSS.MATH.CONTENT.1.OA.D.7	1	Odds & Ends	Treasure Hunt	145
CCSS.MATH.CONTENT.1.OA.D.7	1	Odds & Ends	Human Equations	149
2nd Grade				
CCSS.Math.Content.2.NBT.A.2		Rhythm & Music	The Rhythm Box	113
			Tap, Tap, Clap, Clap	115
CCSS.Math.Content.2.NBT.A.3	Any	Odds & Ends	Place Value Jumping	136
CCSS.Math.Content.2.NBT.B.5	2	Giant 100 grid lessons	Addition with regrouping	34
			Subtraction with regrouping	37
CCSS.Math.Content.2.NBT.B.8	1-2	Giant 100 grid lessons	Add 10	25
CCSS.Math.Content.2.MD.A.3	Any	Yardsticks, Rulers, Tapes	How Long is 100 Feet?	67

3rd Grade

CCSS	Grade	Material	Lesson	Page
CCSS.Math.Content.3.OA.A.1	3	Giant 100 grid lessons	Building the multiplication chart	46
CCSS.Math.Content.3.OA.B.5	3	Giant 100 grid lessons	Building the multiplication chart	46
CCSS.Math.Content.3.OA.C.7	3	Giant 100 grid lessons	Building the multiplication chart	46
CCSS.Math.Content.3.OA.D.9	1-3	Giant 100 grid lessons	Skip Counting Patterns	32
CCSS.Math.Content.3.NF.A.1	3	Balance Beam	Balance Beam Fractions / Measuring Music	103 / 117
CCSS.Math.Content.3.NF.A.2	3	Balance Beam	Balance Beam Fractions / Measuring Music	103 / 117
CCSS.Math.Content.3.NF.A.3	3	Balance Beam	Balance Beam Fractions / Measuring Music	103 / 117
CCSS.Math.Content.3.MD.C.7	3	Yardsticks, Rulers, Tapes	Area: Its all in the Middle Area of a Playground	75
CCSS.Math.Content.3.MD.D.8	3	Giant 100 grid lessons / Yardsticks, Rulers, Tapes	Perimeter / Perimeter: What Goes Around, Comes Around	49 / 69
CCSS.Math.Content.3.G.A.1	3	Odds & Ends	Architecting Polygons	153

4th Grade

CCSS	Grade	Material	Lesson	Page
CCSS.Math.Content.4.NBT.A.2	Any	Odds & Ends	Place Value Jumping	136
CCSS.Math.Content.4.MD.A.1	Any / 4	Yardsticks, Rulers, Tapes	How Long is 100 Feet? / Measuring the School	67 / 78
CCSS.Math.Content.4.MD.A.2	4-5	Yardsticks, Rulers, Tapes	Planning a School Relay	89
CCSS.Math.Content.4.MD.A.3	4	Yardsticks, Rulers, Tapes	Measuring the School	78
CCSS.Math.Content.4.MD.C.5	4	Jump ropes / Odds & Ends	All Turned Around / Parachute Rotation	126 / 156
CCSS.Math.Content.4.G.A.1	4	Jump ropes	Perpendicular or Parallel	124
CCSS.MATH.CONTENT.4.G.A.1	4-5	Projects, Units & Games	Lines, Lines Everywhere	179

5th Grade

CCSS	Grade	Material	Lesson	Page
CCSS.Math.Content.5.NBT.A.1	Any	Odds & Ends	Place Value Jumping	136
CCSS.Math.Content.5.NBT.A.3	Any	Odds & Ends	Place Value Jumping	136
CCSS.Math.Content.5.NBT.B.7	5	Giant 100 grid lessons	Addition with decimals / Subtraction with decimals	40 / 43
CCSS.Math.Content.5.NF.A.2	5	Yardsticks, Rulers, Tapes	Planning a School Relay	89
CCSS.Math.Content.5.NF.B.4	5	Yardsticks, Rulers, Tapes	Planning a School Relay	89
CCSS.Math.Content.5.MD.A.1	5	Yardsticks, Rulers, Tapes	Planning a School Relay	89
CCSS.Math.Content.5.G.A.1	5	Giant 100 grid lessons	Coordinate Graphing	53
CCSS.Math.Content.5.G.A.2	5	Giant 100 grid lessons	Coordinate Graphing	53
CCSS.MATH.CONTENT.5.OA.A.2	5	Odds & Ends	Treasure Hunt	45

6th Grade			
CCSS.Math.Content.6.G.A.4	6	Yardsticks, Rulers, Tapes	Playground Equipment	95
CCSS.Math.Content.6.RP.A.1	6	Odds & Ends	Tangram proportions	162
CCSS.Math.Content.6.RP.A.3	6	Yardsticks, Rulers, Tapes	Build me a Bookshelf	92
			Playground Equipment	95
		Jump ropes	Jumping Contest	132
		Odds & Ends	Paddle Up!	159
	6-8	Projects & Units	Making of a May Pole	189
CCSS.Math.Content.6.NS.C.8	6	Giant 100 grid lessons	Coordinate Graphing	53
CCSS.Math.Content.6.G.A.4	6	Yardsticks, Rulers, Tapes	Build me a Bookshelf	92
CCSS.MATH.CONTENT.6.SP.B.5	6	Projects, Units & Games	School Survey	194
7th Grade			
CCSS.Math.Content.7.G.B.4	7	Odds & Ends	Parachute Circumference	172
CCSS.Math.Content.7.SP.C.5	7	Odds & Ends	Rock, Paper, Scissors	169
CCSS.Math.Content.7.SP.C.6	7	Odds & Ends	Rock, Paper, Scissors	169
CCSS.Math.Content.7.SP.C.7	7	Giant 100 grid lessons	Experimental vs. Theoretical Probability	57
8th Grade			
CCSS.Math.Content.8.G.B.7	6-8	Projects & Units	Making of a Maypole	189

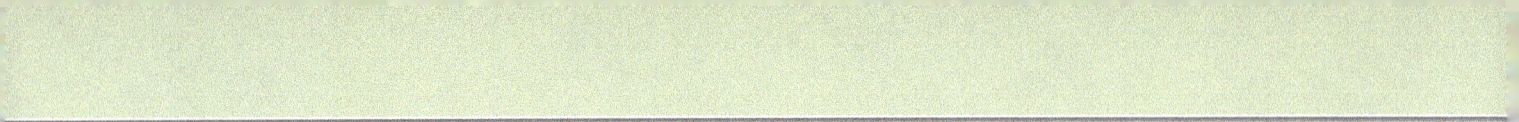

Works Cited

Alpiner, N. (2004). *The role of functional MRI in defining auditory-motor processing networks.* White paper presented at 65th Annual American Physical Medicine and Rehabilitation Conference, Phoenix, AZ..

Baniel, A. (2012). *Kids beyond limits.* New York: Perigee.

Babyak, M., J.A. Blumenthal, S. Herman, P. Khatri, M. Doraiswamy, K. Moore, W. E. Craighead, T.T. Baldwicz, and K. Ranga Krishnan. (2000). Exercise treatment for major depression: Maintenance of therapeutic benefit at 10 months. *Psychosomatic Medicine. 62*:5 p. 633-638.

Bell, R. *Novel Testing Methods and Clinical Applications of Primitive Reflexes* retrieved from http://www.spectrumak.com/resources/spectrum-articles/novel-testing-methods-clinical-applications-of-primitive-reflexes.html on 3/12/2013.

Blomberg, H. & M. Dempsey. (2011). *Movements that Heal.* Queensland: Book pal.

Butterworth, B., and D. Yeo. 2004. *Dyscalculia guidance.* London: David Fulton Publishers

Campbell, D & A. Doman (2012). *Healing at the speed of Sound.* New York: Penguin Group.

Dennison, P. (2006). *Brain Gym and Me.* Ventura, CA: Edu-Kinesthetics, Inc.

De Gacia, L. A. (2014). *Learning is not one size fits all.* Retrieved from Whole Child Learning Solutions at http://www.wholechildlearningsolutions.com

Doidge, N. (2007). *The Brain that Changes itself.* NY: Penguin Group.

Etra, J. (2006). *The Effect of Interactive Metronome Training on Children's SCAN-C Scores.* Applied Dissertation, Nova Southeastern University.

Goddard, S. (2005). *Reflexes, learning and behavior: A window into the child's mind.* Eugene: Fern Ridge Press.

Goddard Blythe, S. (2007). *The well Balanced Child: Movement and learning.* Gloucesterschire, UK: Hawthorn Press.

Goddard Blythe, S. (2009). *Attention, balance, and coordination: The ABC of learning success.* West Sussex, UK: Wiley-Blackwell.

Gold, S. (2002). Exercises to Help your Child, Fern Ridge Press. Retrieved from http://www.fernridgepress.com/autism.exercise.phases.html on 2/12/2013

Gold, S. (2008). *If Children Came with Instruction Sheets.* Eugene: Fern Ridge Press.

Hannaford, C. (1995). *Smart moves: Why learning is not all in your head.* Atlanta: Great Ocean Publishers.

Hannell, G. 2005. *Dyscalculia: Action plans for successful learning in mathematics.* New York: David Fulton

Jordan-Black, J. A. (2005). The effects of the Primary Movement Programme on the academic performance of children attending ordinary primary school. *The journal of Research in Special Education Needs* 5:3 p. 101-111. 11/11/05

Kinoshita, H. (1997). Run for your brain's life. *Brain Work, 7*(1), 8.

Kokot, S. *Integrated Learning Therapy: Unraavelling causes of Learning and Behaviour difficulties* –online video: www.ilt.co.za/video-gallery retrieved on 8/21/13.

Kuhlman, K. & L. J. Schweingart, (1999). *Movement, music and timing: Timing in child development.* Retrieved from High Scope at http://www.highscope.org/Content.asp?ContentId=234 on August 25, 2013.

Lengel, T., & M. Kuczala. (2010). *The Kinesthetic Classroom: Teaching and Learning through Movement.* Thousand Oaks, CA: Corwin.

Lesh, R. A., T. R. Post, and M. J. Behr. 1987. Representations and translations among representations in mathematics learning and problem solving. In *Problems of representation in the teaching and learning of mathematics,* ed. C. Janvier, 33–40. Hillsdale, NJ: Lawrence Erlbaum.

Melillo, R. (2010). *Disconnected Kids.* New York: Perigee

National Center for Learning Disabilities. 2006. Fact sheet: Dyscalculia. Retrieved from http://www.ldonline.org/article/13709/ on October 2007.

O'Dell, N. E. & P. A. Cook. (2004). *Stopping ADHD.* New York: Penguin Group.

Oden, A. (2004). *Ready Bodies Learning Minds.*

Othmer, S. (2012, June 19). *Neurofeedback and the listening program.* Provider teleseminars for Advanced Brain Technologies.

Pheloung, B. (1997). *Help your class to learn.* New South Wales: Griffin Press

Pheloung, B., & J. King. (1993). *Overcoming Learning Difficulties: How you can help a child who finds it hard to learn.* NY: Doubleday.

Pheloung, B. & J. Liljeqvist. (2004). T*en gems of the brain.* New South Wales: Move to Learn.

Ratey, J. J. (2008). *SPARK: The revolutionary new science of exercise and the brain.* New York: Little, Brown and Company.

Sallis, J. F., T. L. McKenzie, B. Kolody, M. Lewis, S. Marshall, & P. Rosengard. (1999). Effects of health-related physical education on academic achievement: Project SPARK. *Research Quarterly for Exercise ad Sport. 70*:2 p. 127-134.

Sousa, D. A. (2008). *How the Brain learns Mathematics.* Thousand Oaks, CA: Corwin.

Story, S. *Getting to the Core of Sensory Issues.* Retrieved from www.moveplaythrive.com on 5/1/2013.

Taylor, M., S. Houghton, & E. Chapman (2004). Primitive Reflexes and Attention-Deficit/Hyperactivity Disorder: Developmental Origins of Classroom Dysfunction. *International Journal of Special Education.* 19:1.

Thaut, M.H., Kenyon, G.P., Hurt, C.P., et al. (2002). Kinematic optimization of spatiotemporal patterns in paretic arm training with stroke patients. *Neuropsychologia*, 40, 1073-1081.

Thayer, R.E. (1996). The origin of everyday moods. NY: Oxford Press.

The brain-movement connection. retrieved 2/5/13 from http://www.skillstrainer.co.k/stnews2/brain2.html

Ullman, M. T. (2005). *Contributions of memory circuits to language: the declarative/procedural model.* Retrieved from *www.ling.hawaii.edu/clrg/contributions_of_memory_11_10_05.pdf* on 7/15/2013.